The Business of Breeding

First published 1990 by Cornell University Press.

International Standard Book Number 0-8014-2233-7
Library of Congress Catalog Card Number 89-46179
Printed in the United States of America
Librarians: Library of Congress cataloging information appears on the last page of the book.

∞The paper used in this publication meets the minimum requirements of the American National Standard for Permanence of Paper for Printed Library Materials Z39.48—1984.

The Business of Breeding

Hybrid Corn in Illinois, 1890–1940

Deborah Fitzgerald

Cornell University Press
Ithaca and London

To the memory of my parents,
Earl Stephen and Sybil Genevieve Fitzgerald

Contents

Illustrations

Acknowledgments

I grew up in a town of twelve hundred people in western Iowa, a farming community that has mirrored the changes occurring in farm towns across the country as agricultural innovations have been introduced. My father was a banker, not a farmer, but both my uncle and my brother-in-law have farmed all their lives, and even my father sold hybrid corn on the side for a few years and farmed a small acreage. Still, I did not enter graduate school eager to study my ancestral homeland; indeed, quite the reverse. But after studying the social history of science and technology and the nonscientific reasons for scientific and technological change, I saw that the history of agricultural development was ripe for analysis.

My debts, both intellectual and personal, span more years than I care to admit and at least three institutional homes. The University of Pennsylvania, Harvard University, and MIT all provided material support of one kind or another that enabled me to finish a leg in this journey. P. Thomas Carroll financed one of my early research trips to Illinois, a kindness that still amazes me. While conducting research I was fortunate to work with several people whose agricultural experience presented questions I had not considered. At the University of Illinois, Maynard Brichford at the University Archives and D. E. Alexander in the Department of Maize Genetics were especially helpful in locating and interpreting documents. At Funk Brothers, Leon Steele and in particular Eugene Funk, Jr., were generous with both their time and their historical materials. Mr. Funk's hospitality, openness, and sense of humor made this part of the research especially enjoyable. At Pfister, Merle Fulton and the late Walter Pfister made helpful suggestions, as did Ron Scherer and Leo Oleson at DeKalb-Pfizer Genetics. Librarians at the National Archives were likewise helpful. Kim Pelkey, Marie Sheehan, and Bobbi Kelley all processed my words with speed and good humor, for which I am grateful.

Others who have helped include A. Richard Crabb, who kindly gave me a copy of his book *The Hybrid Corn Makers*, a major source for this project; Lew Heghn, an independent corn breeder and friend, who answered my technical questions with great patience; and my uncle, Dean Unkel, and brother-in-law, Lowell Aldrich, both Iowa farmers, who answered my practical questions and helped me understand agriculture from a farmer's point of view. Thanks go as well to Margaret Rossiter and an anonymous referee at Cornell, whose criticisms on an earlier draft of this book were both pointed and welcome. I am grateful to Peter Agree and Marilyn Sale at Cornell University Press for providing timely encouragement and reassurance, and to Judith Bailey for her remarkably graceful and intelligent editing.

During my years at the University of Pennsylvania's Department of the History and Sociology of Science, fellow graduate students, faculty, and staff provided both a provocative intellectual climate and a familial sense of community. Charles Rosenberg was a source of great inspiration and support. His own studies of agricultural science provided the intellectual model upon which this book was built, and his interpretive style, which is evident throughout, helped me understand how historical materials resist taming and why ambiguity is so persistent. Barbara Kimmelman, whose friendship saved me more than once from despair that this project might never reach fruition, was as well my toughest critic and most knowledegable judge of ideas and approaches. Thomas Broman, James Capshew, Richard Gillespie, Alex Laszlo, Lynn Nyhart, and Jack Pressman provided much-needed criticism of an early section of this work. Others at Penn who provided support of one kind or another include Robert Kohler, Henrika Kuklick, Judith McGaw, and Ellen Koch.

Colleagues and friends at Harvard were of enormous help both in rethinking the original material and in their general support. I especially thank Everett Mendelsohn, Barbara Gutmann Rosenkrantz, Elizabeth Keeney, and Allan Brandt. Diane Paul, Jean-Pierre Berlan, and Richard Lewontin forced me to reconsider the genetics of hybrid corn, and Diane in particular was encouraging and patient in helping me figure out the science. If overdominance and heterosis do not figure largely in this book, it is not their fault. Joel and Rivka Perlmann's friendship and Joel's computing advice were crucial in the last year especially, and I thank them sincerely.

Parts of the Conclusion originally appeared in Lawrence Busch and William B. Lacy, eds., *The Agricultural Scientific Enterprise* (Boulder, Colo.: Westview Press, 1986). I thank the Press for permission to use the material here.

Finally, my most enduring gratitude is to my husband, Eric Sealine, whose own commitment to clarity and elegance of expression set an example for me and for our son, Jacob, whose life began at the same time as this project. Jake's unfailing lack of interest in this book and his firm belief that there is more to life than words helped keep things in perspective.

DEBORAH FITZGERALD

Cambridge, Massachusetts

The Business of Breeding

Introduction

Just as agricultural historians have neglected the relationship between science, technology, and agriculture, so historians of science and technology have been rather reluctant to examine the origins and development of the agricultural sciences in the United States. Both omissions are predictable. The scholarly bias of agricultural historians and historians of science reflects a general unwillingness to cross the hazardous boundaries of the increasingly specialized subdisciplines of history. Historians of science leave agriculture to historians of agriculture, the latter leave science to historians of science, and the history of agricultural science is claimed by no one. Further, agricultural historians, linked institutionally with sociologists, economists, and American historians, have, for the most part, been professionally inclined to view agricultural development as the product of social, legislative, cultural, and political forces. The role of science and technology in American agriculture, though tacitly recognized as fundamental, has been all but ignored in their studies. For historians of science, agricultural science represents a distinctly "blue collar" phenomenon, which, like engineering, suffers from neglect partly because of its practical aspects. If the physical sciences are highest in intellectual status, the agricultural sciences are near the bottom.[1]

[1]Charles Rosenberg takes exception to this ranking in many essays. Some of his best are collected in *No Other Gods: On Science and American Social Thought*

On the other hand, an agricultural science such as plant breeding *is* a completely different discipline from genetics, not because it is less scientific but because of the more visible uses to which it is put. During the first third of this century, plant breeders and geneticists received virtually the same training, often at the same schools, and most shared a common set of skills, procedures, and professional knowledge. What made them different were their agendas: geneticists focused on explicating the mechanisms of heredity for the use of other geneticists, whereas plant breeders focused on the same thing but for the use of farmers. As others have pointed out, their responsibility to a farm clientele determined the sort of questions plant breeders asked and the sort of solutions they sought.[2]

Yet in many respects the dichotomy is only apparent, reflecting the agricultural scientists' presence in the rural community and their reliance on federal and state funding. Other scientists have performed similar functions, if less blatantly. Physicists working on the Manhattan Project or at NASA, chemists at DuPont, biochemists in the pharmaceutical industry—all have directed their efforts toward goals only remotely identifiable as pure science, but they did so less visibly than agricultural scientists.

It seems to me that the development of agricultural science is particularly interesting precisely because of its ambiguous role as

---

(1978), pp. 135–209. Some of the frustrations, as well as opportunities, for historians of American science are suggested in Rosenberg's "Science in American Society: A Generation of Historical Debate," *Isis*, 1983, 74:356–367. Other important works in agricultural science and technology include Margaret Rossiter, *The Emergence of Agricultural Science: Justus Liebig and the Americans, 1840–1880* (1975); Pete Daniel, *Breaking the Land: The Transformation of Cotton, Tobacco, and Rice Cultures since 1880* (1985); John L. Heitmann, *The Modernization of the Louisiana Sugar Industry, 1830–1910* (1987); Alan Marcus, *Agricultural Science and the Quest for Legitimacy* (1985); and Jack Ralph Kloppenburg, Jr., *First the Seed: The Political Economy of Plant Biotechnology, 1492–2000* (1988).

[2]Rosenberg, "Science, Technology and Economic Growth: The Case of the Agricultural Experiment Station Scientist, 1875–1914," *Agricultural History*, 1971, 45:1–20. In *Science, Agriculture, and the Politics of Research* (1983), Lawrence Busch and William B. Lacy consider agricultural scientists' choices of research problems in a contemporary context.

a scientific discipline engaged in the practical application of scientific knowledge to social and economic problems. The peculiar needs of the rural population—reduced farming costs, stable productivity and efficiency, diverse crop alternatives, broadened political support—have obliged agricultural scientists to consider science itself as a negotiation between natural and human forces. Regional farming realities have dictated the ways in which scientific expertise was used and have provided the experiment station scientist with a range of problems requiring solutions constructed from scientific, economic, and practical considerations.

One way of looking at the production of scientific knowledge in agriculture and the transformation of that knowledge from theory into practice is to examine the development of agricultural commodities. Hybrids such as corn, wheat, sorghum, poultry, and cattle provide vehicles through which to discover how the raw material of plants and animals was modified not only by genetic principles but by economic and social considerations as well. Hybrid corn is a particularly useful subject because in its development the method of creating viable hybrids was invented and refined.

This book has been guided by two questions: first, by what process does "science pure" become "science applied"? and second, how does this process differ between land-grant institutions and private industry? The second question is crucial not only because private enterprise was an important research context in hybrid corn development but also because historians have largely ignored agribusiness as a powerful locus of innovation.[3] Our

[3]An important exception is the history of agricultural implements. See, for example, Wayne G. Broehl, Jr., *John Deere's Company: A History of Deere and Company and Its Times* (1984); E. P. Neufeld, *A Global Corporation: A History of the International Development of Massey Ferguson, Limited* (1969); Walter A. Payne, ed., *Benjamin Holt: The Story of Caterpillar Tractor* (1982); Reynold M. Wik, *Benjamin Holt and Caterpillar Tractors and Combines* (1984); Colin Fraser, *Tractor Pioneer: The Life of Harry Ferguson* (1973); Darragh Aldrich, *The Story of John Deere: A Saga of American Industry* (1942); Robert C. Williams, *Fordson, Farmall, and Poppin' Johnny: A History of the Farm Tractor and Its Impact on*

abundant historiography in the history of ideas, as well as our more modest and imperfect understanding of science in the land-grant universities, lead us to believe that science is at very least an academic exercise (pure), whereas agribusiness is a strictly commercial pursuit that turns existing knowledge to less lofty ends (applied). I have found that the interaction between pure and applied science is neither so simple nor so linear.

To address these issues I focus here on hybrid corn development in Illinois between the late nineteenth century and the early 1940s. Illinois was chosen for several reasons. Most of the seed companies involved in the early research and development of hybrids are located in Illinois—Funk Brothers in Bloomington, Pfister Hybrid Corn Company in El Paso, and DeKalb Agricultural Association (now DeKalb-Pfizer Genetics) in DeKalb. In the interest of discerning patterns of commercial development in the seed industry, it made sense to look at companies whose ecological position was essentially identical. That is, all three companies shared nearly the same opportunities and constraints by virtue of their shared Illinois location. A fourth major company—Pioneer in Iowa—was blessed and cursed (mostly blessed) with an entirely different set of circumstances that suggest a rather different treatment.[4] Illinois was also chosen because corn is the state's leading crop, and thus there existed the economic incentive to pursue corn improvement. The interest in corn spawned a collection of farm groups and university programs designed to support such efforts, and these became the arenas for dispute and resolution of hybrid corn questions.

America (1987); Thomas Norman, *Minneapolis-Moline: A History of Its Formation and Operations* (1976).

[4]Pioneer was started by Henry A. Wallace, whose influence in national agricultural matters was probably unparalleled in the 1930s. As secretary of agriculture from 1933 to 1940, vice-president of the United States from 1941 to 1945, and editor of *Wallaces' Farmer* from 1921 to 1933, he was able to direct both rural sentiment and national policy to a degree that few other individuals could. Further, Pioneer's diversified research program, in which successful hybridizing techniques for corn were turned to animals such as poultry, and the company's close ties to Iowa State University indicate a more broadly synthetic approach than this book can accommodate.

Illinois was in some respects anomalous and in other respects representative of patterns of hybrid corn development as they emerged in different states. The concentration of private companies devoted to research on hybrid corn was unique and created special problems for the university, which, I suspect, would have behaved quite differently in the late 1930s had the private seedhouses been less powerful. Among corn-belt states, however, I think Illinois was more representative than not. Universities in these states developed similar organizations for coping with hybrid corn regulation, production, and distribution and made a deliberate attempt to standardize or at least systematize such procedures from state to state. Because all land-grant universities operated according to the same federal rules and were designed with fairly similar administrative structures, their activities were quite similar as well.[5]

A few words about the scope and organization of this study are in order. First, chronology is not always obvious in the arrangement of chapters. Chapter 1, which describes different attempts to improve the yield of corn through breeding and the shift of many breeders from the selection method to hybridization after Mendel, serves not as a map of what is to come but rather as a lexicon of the technical interests and concerns that informed particular research strategies considered later. Chapter 2 describes the United States Department of Agriculture Bureau of Plant Industry, in charge of the federal corn improvement program. Significant primarily as a clearinghouse for agricultural innovations and consideration of regional differences in farming practice and scientific research, the BPI was also important in setting priorities in the states. By following the crop improvement concerns of bureau personnel, as well as the influence of such breeders as Donald Jones and Henry Wallace on the bureau, one can see the unmistakable shift not only in corn improvement methods but in farming practice as well. Chapter 3, which does

[5]In *The Hybrid Corn Makers: Prophets of Plenty* (1948), Richard Crabb discusses all four major companies, as well as several state and federal research sites.

not consider corn at all, describes the creation and early years of the Illinois Agricultural Experiment Station and the pattern established by its first dean, Eugene Davenport. Here I examine the interaction of federal and state officials in fashioning agricultural college programs that reflected both national and local agricultural interests and created a context within which corn improvement was not only a possible avenue of research but, indeed, a necessary one. This story, essential to an understanding of why the agricultural college and station behaved as they did in the 1920s and 1930s, indicates that the commitments Davenport made to farmers and seed companies in order to ensure their continued support of the college ultimately constrained university research and development options.

Having established the issues of corn improvement and the situation at Illinois, I turn to three case studies to consider how and why traditional selection methods were abandoned in favor of the hybrid method. Chapter 4 examines the particular projects in corn improvement undertaken at the University of Illinois between 1910 and 1935, focusing especially on the relationship between theory and practice. Comparing the work of the Agronomy Department with that of the Cooperative Extension Service, I explore the ways in which current scientific trends were interpreted and transmitted to the farming clientele and, in particular, how research and extension strategies diverged and, in some respects, became incompatible. In Chapter 5 the commercial context of corn improvement is discussed as it developed at Funk Brothers Seed Company, which began its work at nearly the same time as the university. I explore how Funk's determination to devise practical improvements dictated the sort of research questions it asked. In all three situations my questions are similar: what were the aims and methods of traditional crop improvement efforts? how did these change with the invention of the commercially important double-cross hybrid method in 1918? how did the decision to "switch" to hybrids both reflect and alter the relationship between breeders, farmers, and scientists? Finally, in Chapter 6 I discuss how the shift to hybrids created a dramatically different climate for agricultural practice and research. As hybrid corn was transformed from a common research program into a popular and profitable commodity around 1936, the cooperation that had obtained between the university and the seed companies began to disintegrate as their interests and goals diverged.

This book does not attempt to answer all possible questions about either hybrid corn or what might be called the land-grant/agribusiness complex. In recent years some scholars have begun considering hybrid corn in a broader context; Jack Kloppenburg has discussed the role of hybrids in the emergence of biotechnology and the recurring interest in plant patents; Jean-Pierre Berlan and Richard Lewontin have raised important questions about the scientific validity of hybrids and the heterosis concept; and of course the controversial relationship between the land-grant system and agribusiness was most powerfully addressed nearly two decades ago by Jim Hightower.[6] Although the issues raised by these and other scholars have informed my work in important respects, my concerns are more deeply grounded in historical reconstruction. This is ultimately a case study built from historical scraps, and its aim has been to discuss in a particular way and for a particular era how hybrid corn came to exist, how the agricultural college got involved in commodification, and how the research-and-development interests of agribusiness have changed the way the agricultural college conceives of its mission. Some of my conclusions are patently conjectural, and some subjects are treated more briefly than they deserve. I will leave it to others to wrestle with the tigers hinted at here.

Finally, although in no way definitive, this book does, I hope, establish the terrain upon which issues of scientific agriculture and agricultural science were recognized and considered. As

[6]Kloppenburg, *First the Seed*; Jean-Pierre Berlan and Richard Lewontin, "The Political Economy of Hybrid Corn," *Monthly Review*, 1986, 38:35–47; Jim Hightower, *Hard Tomatoes, Hard Times—the Original Hightower Report, Unexpurgated, of the Agribusiness Accountability Project on the Failure of America's Land Grant College Complex* (1972).

agriculture is a negotiation between science and practice, no study of agricultural science can be complete without a consideration of the contextual factors that pervaded the agricultural scientist's agenda. Science was rarely something that was "done to" farmers; rather, it developed as a compromise between theory and necessity. Negotiations among farmers, seed companies, and university breeders most often occurred in the realm of assumptions each had about the best interests of the other. To interpret these assumptions and the activities they engendered is what this history is ultimately about.

1

Corn Breeding in Theory and Practice

There is no evidence that farmers and breeders were ever really satisfied with the corn they grew. In some years the corn was better than in other years, but the yearly changes in weather, invasions of pests and disease, and rural prosperity seemed to change the rules of the game with each new planting season. Following a drought, for example, farmers might look for corn that could withstand such conditions; a chinch bug invasion the next year might well change their priorities. Farmers could not afford to ignore poor quality and yield; if they could squeeze a few more bushels out of each acre, their prosperity would increase. This simple logic sustained a great many corn improvement attempts by farmers and seedsmen alike.

One of the most notable features of nineteenth-century corn improvement strategies was their basically democratic nature. As Henry A. Wallace would later remark, "Anybody can cross corn."[1] If this was true of crossing varieties, it was even more true of selection, a method so commonsensical that ordinary farmers with little scientific inclination could practice it as a matter of course. Crossing varieties was a bit more complicated, and thus less widely practiced, but nonetheless constituted a rather commonsense, comprehensible approach to corn improvement. Of

[1] Henry A. Wallace, "The Corn Breeding Plot," *Wallaces' Farmer*, 29 March 1918, 578.

course, not all farmers tried either crossing or selection, and in fact probably only the more affluent ones considered such practices, but the obstacles to doing so were personal rather than scientific.

With the introduction of the Mendelian approach around 1900, this situation changed dramatically. Whereas farmers could still practice corn improvement if they wished, seed producers and breeders were gradually joining the ranks of an emerging scientific group that espoused the apparently more scientific Mendelian approach. Over the next twenty years, such simple methods as selection and varietal crossing would give way to inbreeding and crossing, and the change in method itself would change the social organization of agriculture. No longer would farmers use their experience and expertise to establish and maintain their own high-yielding strains of corn; instead, plant breeders would become the new experts not only on which corn lines were better but on how to create them.

Here it is enough to discuss the alternatives among which farmers and breeders could choose before the mid-1910s. The change in breeding methods was neither so dramatic nor so complete as some have suggested, and as later chapters will show, some breeders maintained their commitment to traditional methods long after their colleagues abandoned such "old-fashioned" techniques.

Early Corn Improvement Methods

One of the reasons why breeders have been attracted to corn, aside from its obvious economic value, is that it is perhaps the most easily manipulated of farm crops. A single corn plant carries both male and female flowers: the pollen is produced by the tassel on the top of the stalk and the seed is produced on the ears by means of the silks. At pollination, the pollen is blown about by the wind, settling randomly on the silks of its own plant or others in the field. This event, called open-pollination, occurs without human intervention and is the traditional, or prebreeding, manner of producing a corn crop.

The selection, or mass selection, method of corn improvement rests on the notion that a little intervention can reduce poor corn and increase good corn. Selection cannot be traced to an originator but was widely practiced throughout corn-growing areas of the United States in the nineteenth century, as it had been for many centuries before that. In this method corn growers selected the most promising ears of corn from the field or corncrib at harvest for planting the following spring. Selection was based primarily on the physical appearance of the ears rather than the productiveness of the plant, two factors that were not necessarily related. If growers believed that smooth kernels or long ears were favorable characteristics in corn and, more to the point, correlated with high yield, they selected such ears. The following spring the farmer mixed the seed from these ears all together and planted the composite. With any luck the seed would very gradually improve to the point that it outyielded the original sample.[2]

As a method of crop improvement, selection had many virtues. It required no special equipment, no scientific training, and no capital investment. It could be practiced by farmers and breeders alike because it required only an experienced eye for productive corn. Further, the rules for selecting were highly flexible, and each grower decided what characteristics to select for and what to select against. Smooth vs. rough kernels, tall vs. short stalks, early vs. late maturing, single-ear vs. multiple-ear plants—there was seemingly no limit to corn's intrinsic variability. For many farmers, the ideal type for which they were selecting possessed the characteristics codified in the corn show scorecard. The score-

[2]The most lucid account of early breeding methods is in George F. Sprague, *Corn and Corn Improvement* (1954), pp. 221–225, 229. See also Norman Simmonds, *Principles of Crop Improvement* (1979). Other grains, such as wheat, oats, and horticultural plants, had been studied abroad on both theoretical and practical grounds; for an account of this work in England, Germany, and Sweden, see Hugo De Vries, *Plant Breeding* (1907), pp. 1–106.

card, introduced in many corn-growing states in the 1890s, was a list of features that corn show judges thought prize-winning corn should have. In Illinois this list was drawn up by the Illinois Corn Breeders' Association; it included preferences for ear length and circumference, rough kernels, and general "fancy" appearance. For some farmers, no doubt, what was good enough for the corn show was good enough for them, and they dutifully selected their seed corn accordingly.[3]

To some growers, however, most notably those with some scientific training, selection appeared to be of limited value in improving corn yield. First, they noted, most growers were selecting purely on the basis of appearance, which may or may not correlate with high yield. Many growers might even be selecting against high yield by choosing ears for aesthetics rather than productivity. Second, and much more serious from their point of view, the growers were selecting only the female parent, that is, the seed ear.

Once the selected seed was planted, growers usually made no attempt to control pollination, figuring that if all the seed was good, then open pollination was satisfactory. Thus, at harvest growers could not determine which pollen-bearing plants had fertilized the ears and were unable to ascertain whether the particular ear characteristics were contributed by the male or female parent. In general, critics felt, mass selection was an inefficient way to improve a field, rather than a strain, of corn and worked on such a highly mixed collection of characters that selecting for any particular character could result in only modest improvement.

One of the earliest critics of selection was W. J. Beal of the Michigan Agricultural College. In the late 1870s, Beal described his experiments in crossbreeding corn varieties to increase yield, which for the first time considered the male as well as the female parent. Drawing on his knowledge of animal breeding, as so many

[3]For a brief discussion of "show corn," see Helen Cavenagh, *Seed, Soil, and Science: The Story of Eugene D. Funk* (1959), pp. 216–219. This issue will be discussed more fully later.

early breeders were to do, Beal criticized plant breeders for ignoring the male parent: "What do we think of a man who selects the best calves, pigs, and lambs from the best mothers, paying no attention whatever to the selection of a good male parent? This is what our very best farmers are doing all the time with their seeds and plants." If nothing else, Beal declared, breeders using mass selection should detassel all corn plants exhibiting unfavorable characteristics to prevent them from passing these characters on to the nascent ears.[4]

Beal's most notable contribution to breeding technique was his method of varietal crossing, which consisted of selecting two strains of corn showing favorable characteristics and planting them in alternate rows. As the tassels began to appear, Beal removed them on plants in every other row, thus making certain that all detasseled plants had been fertilized by the tasseled plants and, of course, that the tasseled plants had been self-fertilized. Beal was thus able to control the pedigree of each cross.

This method was important for several reasons. First, by manipulating both male and female parents, the breeder was able to repeat a particular cross and, in a small measure, predict its performance. This repeatability was potentially useful for commercial breeders who wanted to retain, or "fix," favorable strains. Second, Beal's varietal crosses outyielded ordinary open-pollinated varieties by 10 to 50 percent, an improvement that most other breeders could not afford to ignore. And finally, Beal's research inspired experiment station scientists to investigate his claims further. In the 1880s many, if not most, midwestern stations began using varietal crossing as a means of improving their own varieties and yields.[5]

It appears, however, that varietal crossing was not used very extensively by commercial breeders. Results from early experi-

[4]William J. Beal, "Report for 1876," Michigan State Board of Agriculture, pp. 213–214; see also W. Ralph Singleton, "Early Researches in Maize Genetics," *Journal of Heredity*, 1935, 26:50.

[5]Sprague, *Corn Improvement*, p. 226.

ments were considered inconsistent, and most breeders felt that it was not worthwhile to change from the straightforward, traditional method of selecting promising seed and planting it en masse to a complicated, unproven technique that required segregating the strains being used in the cross as well as detasseling half the plants.[6]

One of those inspired by Beal's work was George E. Morrow, who began crossbreeding experiments at the Illinois experiment station in 1888. Although Morrow's chief interest was in field methods rather than breeding, the experiments he conducted through 1893 also included breeding as one of eleven factors to consider in improving the yield of corn. He followed Beal's method and reported an average yield increase of 9.5 bushels per acre. His brief discussion of crossing was an unqualified endorsement of the practice.[7]

In 1889 George W. McCleur joined the corn project as assistant station horticulturalist. Like many other nineteenth-century agricultural scientists, McCleur was nearly overwhelmed by the immense ignorance of commercial breeders regarding breeding methods and results. Noting wryly that "breeders of corn are not prolific writers and so are not well known," McCleur observed that a few farmers had experimented with crossing varieties but that the results had not been as rewarding as those obtained by "continued careful selection." But like Beal, McCleur was convinced that if crossbreeding could improve flowers and livestock, it should just as surely improve agricultural cereals and grains.[8]

[6]For example, whereas *Bulletin* 21 of the Illinois Agricultural Experiment Station concluded that varietal crosses were superior to open-pollinated varieties, the Kansas station found the evidence from its studies inconclusive. See *Office of Experiment Stations Bulletin* 15, 1893, 81–82; Robert W. Jugenheimer, *Hybrid Maize Breeding and Seed Production* (1958), p. 6.

[7]George Morrow and Frank Gardner, "Field Experiments with Corn," *Illinois Agricultural Experiment Station Bulletin* 24, 1893, 173; see also George Morrow and Thomas Hunt, "Field Experiments with Corn," *Illinois AES Bulletin* 4, 1889, and *Illinois AES Bulletin* 13, 1891. The other ten factors were depth of planting, depth of cultivation, frequency of cultivation, time of planting, density of planting, effect of root pruning, growth of corn plant, planting hills or drills, effect of removing tassels, and tests of different varieties.

[8]George McCleur, "Corn Crossing," *Illinois AES Bulletin* 21, 1892, 82.

McCleur's experiments consisted of selecting twelve common varieties of corn and crossing them in various configurations, inbreeding some varieties. His method was more exacting than Beal's and, in fact, was generally adopted by later breeders in their hybrid studies. To control pollination, McCleur placed cloth bags over both the tassels and the silks to prevent random pollination. When the silks were ready for pollination, he and his assistants would "gather the pollen on a sheet of smooth paper and roll it up funnel shaped. Next raise an umbrella and hold it in such a way as to keep all flying pollen from the ear, remove the bag, and apply the pollen until the silks are almost hidden." McCleur's standard for measuring the degree of improvement was the weight of a ten-ear sample, which he compared with the weight of the original variety and sometimes the inbred variety as well. For example, if the two strains being tested were A and B, he would weigh ten ears of A × B, ten ears of A and/or B, and ten ears of A or B that had been self-pollinated.[9]

His conclusions—more properly, his observations—were somewhat ambivalent. While he found that different strains are easily cross-pollinated, it was not clear that they were much superior to the original varieties, for they tended to revert to inferior types: "Two varieties that have been long selected for opposite or widely different qualities must when crossed tend to neutralize most strongly the very trait which we have with so much pains brought out and maintained."[10] Furthermore, McCleur discovered that he could not predict which ears or varieties would yield more. His original theory, that closely related strains would improve more dramatically than distantly related strains when crossed, proved to be unfounded. Unfortunately, the converse—that distant relatives when crossed would improve more—was likewise not entirely true. Rather, there seemed to be no obvious explanation why some crosses worked and others did not; it was simply a matter of chance. And finally, McCleur's work seemingly reaffirmed the age-old belief that inbreeding was a

[9]Ibid., p. 100. Later breeders dispensed with the umbrella, instead simply attaching the pollen bag securely to the maturing ear.

[10]Ibid., p. 99.

dead end in plant improvement. Like his contemporaries, he found that self-pollinating corn produced runty, misshapen, frequently barren ears. These he did not attempt to cross. Indeed, if the goal of breeding was to increase the vigor and yield of a strain and if visible characteristics provided the basis for selecting one strain over another, then crossbreeding inbreds would merely satisfy a breeder's perverse curiosity; it could not be expected to do more. Yet McCleur was not entirely discouraged by his mixed results. His aims, like his findings, were modest, and he appeared satisfied that a beginning had at last been made, saying, "If the man with false ideas as to plant breeding can succeed in making improvements, the man with correct notions should be so much the more successful."[11]

Varietal crossing was practiced periodically well into the twentieth century, and in 1906 Secretary of Agriculture James Wilson enthusiastically praised strains bred in Ohio, which increased yields more than ten bushels per acre.[12] Nonetheless, the generally mixed results growers obtained in crossing varieties usually produced more restrained reactions. In the same year, J. I. Schulte of the Office of Experiment Stations, noting that varietal crossing was being investigated at some experiment stations, reported that selection was still the best method. This claim he supported with experimental evidence from Wisconsin, where one strain more than doubled its weight after selection. And in 1908 the USDA reported that in some areas selection alone had increased yield per acre to between ninety and a hundred bushels more than the average yield.[13]

[11]Ibid. Corn geneticist D. E. Alexander differs with this interpretation of McCleur's work, lamenting that McCleur "failed to grasp the practical significance of the vigor he observed in crosses of mildly inbred strains." See Alexander, "Illinois and the Beginnings of Hybrid Corn" (manuscript, 1963, Alexander Collection, Department of Maize Genetics, University of Illinois).

[12]"Report of the Secretary," *Yearbook of Agriculture*, 1906, 52.

[13]J. I. Schulte, "Corn Breeding Work at Experiment Stations," *Yearbook of Agriculture*, 1906, 292–293; "Report of the Secretary," *Yearbook of Agriculture*, 1908, 53.

Corn Breeding at Illinois, 1895–1910

As dean of the University of Illinois College of Agriculture, Eugene Davenport was an important supporter of the corn-breeding work throughout his tenure. Although his scientific expertise concerned livestock, rather than corn, breeding livestock according to their pedigrees had become a standard technique, and the analogous breeding of crop plants seemed a fruitful direction for breeding methods to take. In addition, Davenport had worked briefly with Beal while at Michigan and had observed the substantial increases in yield and quality systematic breeding could produce. The persuasive Davenport made his interests known to the trustees, who in 1895 adopted a resolution that "Indian corn and its relations from every conceivable point of view ought to be considered the foremost subject for experimentation at this station."[14]

During the following years Davenport hired four young agricultural scientists, who kept the corn-breeding projects alive. In 1896 he appointed Perry Greely Holden, his former assistant at Michigan, as assistant professor of agricultural physics. Two years later he appointed Archibald Dixon Shamel instructor in farm crops and manager of the experiment station after Shamel received his B.S. at Illinois. In 1900 Davenport named Edward Murray East assistant chemist at the station after East received his B.S. and three years later promoted him to assistant plant breeder. And finally, in 1904 Davenport hired H. H. Love, who had just completed a chemistry B.S. at Illinois Wesleyan. Although their time at Illinois was brief, all four men were introduced to breeding there and later became prominent in the field.[15]

Under Davenport's guidance the corn-breeding work, includ-

[14]Minutes of the meeting, 4 September 1895, *Proceedings of the University of Illinois Board of Trustees*, pp. 167–168.

[15]Crabb, *Hybrid Corn Makers*, pp. 19–35 and passim; Richard Gordon Moores, *Fields of Rich Toil: History of the University of Illinois College of Agriculture* (1970), passim. Additional biographical information was taken from *American Men and Women of Science*.

ing inbreeding, continued at Illinois. Although in 1896 Davenport reported that "experiments this year show disastrous results from self-fertilization of corn," two years later the work "assumed gigantic proportions" under Perry Holden. Davenport's assessment of the project in 1898 is worth quoting at length:

> The effects of inbreeding appear both pronounced and disastrous, the second generation from inbred seed being less than two-thirds normal size and nearly barren. Remarkable variations have been brought to light in all portions of the plant, which seems particularly variable, and yet there is a pronounced tendency to respond to selection. The first year, after all imperfect stalks and all extremely large, early or late ones had been removed, not one-fourth of the crop was left, but the second planting from this seed when closely selected after the same plan left almost a full stand, which shows that corn may be brought much nearer a constant type than has ever yet been done.[16]

But while Davenport successfully maintained the studies of varietal crossing, selection, and inbreeding, this work was overshadowed at Illinois by Cyril G. Hopkins's studies of corn chemistry. Hopkins, who received an M.S. in chemistry from Cornell University in 1894, was appointed station chemist six weeks before Davenport began his deanship at Illinois. Following his thesis, "The Chemistry of the Corn Kernel," Hopkins in 1896 began a series of experiments that are today considered classic and, according to some, invented the ear-to-row method of selection and improvement. In these experiments, Hopkins by selection attempted to alter two characteristics—protein and oil content—to the extremes of high and low possible in the kernel. Initially he selected ears that displayed physical characteristics typically associated with one of four factors: high oil, low oil, high protein, and low protein. Once selected, a portion of each

[16]Eugene Davenport, Minutes of the meeting, 8 December 1896, 8 March 1898, *Proceedings of the University of Illinois Board of Trustees*, pp. 42, 224–225.

ear of corn was planted in a row, each row representing one ear. In this way Hopkins could trace the parentage of the plants to the appropriate mother ear and could check the performance of one ear against another.[17] This was the critical improvement in breeding method. With mass selection, progeny could not be checked against the parent because seed was planted en masse. The essence of the ear-to-row was in this progeny test, which in a sense brought corn breeding even closer to the pedigree breeding practiced by livestock experts.[18]

Hopkins first reported his results in 1899, and they were impressive. By means of selection he had increased the average protein content from 10.92 to 12.32 percent and oil from 4.7 to 6.12 percent. This was the first definitive proof that the chemistry of corn could be altered in a particular direction.[19] Not surprisingly, this work absorbed most of the research staff and resources of the station. East, Shamel, and Love all worked on Hopkins's ear-to-row project; East and Love were hired explicitly for the job. In addition, Louis Henrie Smith worked as Hopkins's assistant while obtaining his M.S. in chemistry at Illinois. Holden seems also to have been involved in some way. All Hopkins's assistants were interested in breeding as well as selection, and East especially came to believe that breeding held the greater potential.[20]

Hopkins was optimistic that his selection method would dra-

[17]Hopkins's work is discussed briefly in Sprague, *Corn Improvement*, p. 228.

[18]See Cyril G. Hopkins, "The Chemistry of the Corn Kernel," *Illinois AES Bulletin* 53 (1898); these experiments are also discussed in *Illinois AES Bulletins* 55, *82*, *87*, *100*, 119, and 128.

[19]According to Holden, Hopkins was so pleased with his results that he publicly announced that "he had accomplished the well-nigh impossible, the miracle of doubling the feeding value of King Corn." This exaggerated claim reportedly so upset Davenport that Hopkins was shuttled off to Germany for a year starting in September 1899. But when Holden left the Agronomy Department in 1900, Hopkins replaced him as professor of agronomy. See Holden to Richard Crabb, 21 June 1948, Alexander Collection; Davenport to Hopkins, 25 April 1900, Davenport, Personal Letterbook (8/1/21, University of Illinois Archives [UIA]), pp. 13–14.

[20]Moores, *Fields of Rich Toil*, pp. 150–162; Crabb, *Hybrid Corn Makers*, pp. 14–38.

matically improve corn quality in Illinois. Of particular interest to him was the possibility of developing corn with a much higher percentage of protein and a correspondingly low percentage of oil. He envisioned such corn as a boon to the livestock industry, and in fact, Davenport had early suggested that corn should be designed specifically for livestock needs. Hopkins's studies seemed to indicate that such an idea was viable. Further, he felt that when protein content was increased, the overall crop yield would increase as well. In 1902 the future of ear-to-row selection seemed promising.[21]

In 1900, after Hopkins returned to Illinois from a year in Germany, his corn project gained momentum. In the next few years, it extended throughout the state. In 1901 he made arrangements with area farmers to grow and select corn for the oil and protein experiments, and this strategy significantly expanded the scope of the project. He also enlisted the cooperation of Eugene D. Funk, who, according to Hopkins, had "set up a seed company with the direct purpose of following up on a large commercial scale the work which I have been doing in the matter of breeding seed corn for improvement in quality." While Hopkins may have overstated the case, Funk was in fact an enthusiastic supporter of ear-to-row selection, and he and Hopkins shared information and expertise. If anything, Funk was even more intent than Hopkins on developing specialized corn, and in 1902 he bred strains for high oil and for high protein and low oil content for feeding fancy cattle and young hogs.[22]

As the years passed, however, results from Hopkins's long-term project told a different story. It was indeed possible to alter the

[21]See Hopkins to James Reid, 14 January 1902, Hopkins to C. D. Smith (director of the Michigan Agricultural Experiment Station), n.d. [ca. January 1902], Hopkins to Senator T. S. Chapman, 19 February 1903, Alexander Collection; Crabb, *Hybrid Corn Makers*, p. 15.

[22]Hopkins to J. C. Reid, W. H. Conibear, F. A. Warner, A. P. Grout, J. O. Toland, and E. E. Chester, 17 September 1901, Hopkins to H. B. Gurler, 14 January 1902, Hopkins to Holden (at Funk Brothers), 15 February 1902, Alexander Collection; see also Cavenagh, *Seed, Soil, and Science*, p. 93.

protein and oil content of corn significantly, but the price for this bonus was paid in yield. By 1918 it seemed clear that the ear-to-row method was not the simple answer to increasing corn yield. Commercial seedsmen found that there was really no appreciable market for highly specialized corn, or at least none that warranted the complicated field and laboratory procedures. Supporters of the method, admitting that it was not, after all, a new means of improving corn, praised the experiments for contributing to "our knowledge of the effect of selection on yield."[23]

There seems little doubt that Hopkins's project squeezed the breeding work farther and farther out of the spotlight at Illinois and that breeding continued in a small way in spite of, rather than because of, Hopkins's work. Both Holden and Shamel were bitter in their later years, claiming that Hopkins had never been involved in the breeding work. Holden suggested that Hopkins's indifference to the breeding was partly professional, quoting him as saying, "I am not a plant breeder, horticulturalist, or farmer. I am a chemist!"[24] Shamel's memory was even less kind; he recalled how Hopkins "always called me into his gloomy basement laboratory to vent his spleen. He lectured me very pompously from time to time on the errors of my ways in 'playing' as he expressed it without his official consent. It is also clear that both Hopkins and Eugene Davenport were violently opposed understandably to my work along that and other lines and made it so miserable and unpleasant for me that I decided to leave there."[25] It has been reported that East, too, had difficulty breeding corn while in Hopkins's employ. When East requested permission to

[23]Alexander, "Early Work in Corn Experimentation at Illinois" (manuscript, 1959), Holden to Crabb, 21 June 1948, Alexander Collection; University of Illinois, Department of Agronomy, "Historical Data for President Kinley," February 1941, 8/6/2, UIA, pp. 25–26.

[24]Holden to Crabb, 21 June 1948, Alexander Collection. There was no love lost between Hopkins and Holden. Hopkins, for his part, thought Holden "dishonest" and spoke freely to Davenport of their animosity; see Hopkins to Davenport, 13 October 1903, Alexander Collection.

[25]Shamel to Holden, ca. 14 April 1948, 8/1/51, box 1, UIA.

start an inbreeding study, Hopkins is alleged to have responded, "We know what inbreeding does and I do not propose to spend people's money to learn how to reduce corn yields."[26]

Such stories, derived from memory and possibly embellished for posterity, may be inaccurate or entirely apocryphal, of course, but one suspects that they at least reflect the common disenchantment with corn studies at Illinois. Of the five young breeders, only Smith remained at Illinois past 1905; the rest simply passed through on their way to more breeding-related positions. Holden resigned in 1900 to become manager of a beet-refining plant in Pekin, Illinois; not coincidentally, beets at that time were considered most responsive of all plants to selective breeding. In France, Henry Vilmorin had produced dramatic and widely accepted results in his "creation" of a beet with an unprecedentedly high sugar content. Holden then spent a year helping Eugene Funk establish his seed-corn business before moving to Iowa State University, where he became known as a corn evangelist. Shamel was next to leave, in 1902, moving to the USDA, where he soon became principal physiologist. East was lured away from Illinois by E. H. Jenkins at the Connecticut station and, after publishing his remarkable studies on corn breeding, took a position at Harvard's Bussey Institution in experimental morphology.[27]

By 1905 corn breeding at Illinois had entered a dormant phase. Those interested in breeding had mostly left the university and, despite Davenport's continued enthusiasm for breeding, the focus of corn work settled firmly on Hopkins's chemical studies. University scientists contributed very little to the revolution in plant breeding between 1900 and 1917; not until the 1920s did a new group of breeders produce innovative research that was part of the larger national pattern of Mendelian studies. Ironically, it would be Hopkins's assistant, E. M. East, who launched the first volley.

[26]As told in Grant Cannon, ed., *Farm Quarterly Great Men of Modern Agriculture* (1963), pp. 221–222.

[27]E. H. Jenkins to East, 11 and 31 July 1905, Alexander Collection.

The Challenge of Mendelism

In recent years, much has been made of the profound effect Mendel's work had on corn breeders after 1900, and indeed, in hindsight it has seemed to many that Mendelism definitively altered the practice of corn breeding. But to many breeders well into the 1920s Mendelism seemed to offer little in the way of practical crop improvement. If we have come to think of Mendelism as the long-awaited and nearly immediate solution to corn problems, it is because Mendelism "won" the debate over breeding methods and not because there was no contest. As history repeatedly shows, the triumphant paradigm tends to obscure the previously tenable alternative paradigms, and proponents of the new "truth" seem to lapse into a sort of amnesia regarding the "loser's" previous merits.[28]

While there is little doubt that Mendelism "won" and that it generated an enormous amount of experimentation shortly after the turn of the century, it is important to realize that the alternative breeding methods—selection and varietal crossing—continued to be popular and reasonable corn improvement methods well into the 1920s. Especially among practical corn breeders within the state and federal system, Mendelism seemed of little help to those who were simply trying to increase corn yields. As far as Mendelians were concerned, the selection method in particular had to be ousted; for breeders using selection, the superiority of Mendelism had to be demonstrated to justify abandoning a promising crop improvement method. That Mendelism ultimately won had to do less with its intrinsic superiorities than with its appeal to the first generation of geneticists—and to seed producers who knew a profitable innovation when they saw one.

[28]Thomas Kuhn, *The Structure of Scientific Revolutions* (1962). I am using Kuhn's model rather loosely, for I believe that the paradigm shift in this case resulted less from scientific problems with the old paradigm and the scientific efficacy of the new than from the economic, social, and professional interests of the hybridizers.

The Business of Breeding

Since Darwin's classic study of cross- and self-fertilization in 1876, scientists—particularly those working in Britain and Europe—had been struggling to explain patterns of inheritance by studying and manipulating variation in plants and animals. Many were also motivated by a desire to improve economically important field crops. In Great Britain, E. F. Hallet and Patrick Shirreff improved wheat varieties by carefully selecting individual, generally deviant variations that were superior to the average type. German breeders, rather than search out anomalies or focus on individual plants, sought to improve cereals by selecting a "crowd" of individuals, all exhibiting the superior characteristic. Generally the work of these British and German breeders was guided by commercial concerns, and their methods of improvement, while satisfying their customers, did not contribute much to the theoretical study of inheritance.[29]

At the experiment station in Svaloff, Sweden, however, breeders using the German method of gradual improvement by selection concluded that it was not uniformly reliable. Although mass selection could slowly improve a favorable characteristic, it could not get rid of unfavorable characters expressed in the same plant. In the 1890s Hjalmar Nilsson, director of the station, began a series of experiments in which wheat plants with different variations were selected and then mixed together according to variation type. Plants within each group were nearly identical to each other but different in some way from plants in other groups. One of these groups was very small because the variation it represented was rare. Station workers carefully recorded the parentage of the plants before they were mixed. The following fall, when the progeny were examined, Nilsson discovered that the small rare group had bred true to its parent, and after consulting the records, found that all the individuals had come from a single parent.[30]

[29]Charles Darwin, *The Effects of Cross and Self Fertilization in the Vegetable Kingdom* (1876). Virtually all breeders began their reports with a bow to Darwin. See also Hugo De Vries, *Plant Breeding*. Essays in this book were based on a series of lectures De Vries presented during his American tour in 1906.

[30]De Vries, *Plant Breeding*, pp. 48–106.

This breeding "accident" formed the basis of Nilsson's theory of biotypes. Nilsson hypothesized that individual plants were composed of numerous elemental factors, or biotypes, that roughly corresponded with observable characteristics. Different wheat plants, for instance, were composed of different biotypes within the range of possible variation. By paying careful attention to observable variation among such plants, the breeder could select and isolate favorable biotypes. A similar theory was developed by Wilhelm Johannsen in the early twentieth century. Johannsen, who in 1909 coined the terms "gene," "genotype," and "phenotype," used the notion of pure lines to describe the process of inbreeding to isolate biotypes.[31]

Yet another strand was contributed by Hugo De Vries, whose *Intracellulare Pangenesis* (1899) and *Die Mutationstheorie* (1901–1903) provided an alternative to natural selection. According to De Vries, evolutionary change occurred not just by gradual, continuous variation and selection as Darwin had thought but also by jumps or mutations. These mutations, which arose spontaneously in plants and exhibited a discontinuity with the parent form, could then be selected for and isolated as new types. De Vries based his ideas of variation and inheritance on the principle of unit characters, or pangenes, hereditary units analogous, De Vries postulated, to molecules. As molecules formed the building blocks of chemical elements, so pangenes were the building blocks of plant and animal structure.[32]

It is difficult, however, to evaluate the impact of such theoretical formulations on the work of American corn breeders. Very few seem to have used De Vriesian principles to study inheri-

[31]Frederick B. Churchill, "Wilhelm Johannsen and the Genotype Concept," *Journal of the History of Biology*, 1974, 7:5–30.

[32]De Vries, *The Mutation Theory*, trans. J. C. Farmer and A. D. Darbishire (1910); De Vries, *Intracellular Pangenesis*, trans. C. S. Jaeger (1910). Garland Allen, *Thomas Hunt Morgan* (1978), pp. 97–153; Allen, "Hugo De Vries and the Reception of the Mutation Theory," *Journal of the History of Biology*, 1969, 2:55–87; Barbara Kimmelman, "Hugo De Vries and the Experimental Method" (manuscript, 1978).

tance, and until De Vries's American lecture tour in 1906, which introduced the British and European studies to American breeders, it seems unlikely that Americans knew or cared much about developments abroad. For one thing, the Europeans were studying naturally self-fertilized plants, such as wheat and oats, and were thus able to avoid the powerful Darwinian bias against inbreeding that guided the Americans studying corn. The applicability of European findings to American grains was, at best, unclear. Further, by the turn of the century plant and animal breeders were not always speaking the same language as plant physiologists, cytologists, and evolutionists. For the latter group, and experimental biologists in general, breeding provided a means of studying evolutionary laws and patterns of inheritance. For the breeders, such inquiries were of marginal importance as a rule. They might help explain why inheritance worked, but what the breeders wanted was more specific information on how to direct inheritance toward a marketable goal.[33]

Of considerably more interest to breeders were the theories developed by Gregor Mendel and British biometrician Francis Galton. Although Mendelism and biometry were thought incommensurable in Britain, in America they both enjoyed popularity among breeders because both provided methods of predicting which characteristics would be inherited. Whereas Galton's work was based in statistics rather than biology, it provided a useful guideline for breeders eager to predict the gross features of a varietal cross. The law of regression, for example, which predicted that the progeny of parents who were above or below the average tended to regress toward the average for the characteristic, rang true for breeders interested in such characteristics as plant height, ear length, and so on. It also closely resembled the pedigree method of livestock breeding, in which predictive con-

[33]Breeders were fond of quoting the Knight-Darwin law that stated, "Nature abhors perpetual self-fertilization." For a discussion of the difficulty of using De Vries's work in breeding, see Barbara Kimmelman, "The American Breeders' Association: Genetics and Eugenics in an Agricultural Context," *Social Studies of Science*, 1983, 13:172–173.

trol was gained from particular genealogical information. But unlike animal pedigree breeding, biometry was most successful with large populations rather than individuals, and it was therefore popular among breeders using mass selection.[34]

Mendel's laws, on the other hand, were generalized from the behavior of individual plants. The law of dominance and recessiveness, which stated that a character is either dominant or recessive to its alternative character, such as wrinkles and smoothness in peas, was a straightforward and tangible rule that breeders were quick to adopt. Mendel's law of probability, which predicted the likelihood of particular inheritance patterns, was likewise examined and quickly approved. Like other biologists, many practical breeders had observed tendencies toward inheritance patterns without being able to identify or generalize the pattern itself. For them the Mendelian laws rang true at once; they had only to apply them to the specific case of corn.[35]

For the more theoretical biologists, too, Mendel's laws provided a research agenda that gathered the divergent threads of evolution, development, and inheritance. Largely in response to the growing excitement over recent studies of inheritance, of which Mendel's rediscovered work loomed the largest, Willet M. Hays, then assistant secretary of agriculture, took the lead in establishing the American Breeders' Association in 1903. Under the banner of "breeding," Hays collected a group remarkable for its diversity yet reflective of both the subject's ambiguity and the loose organization of turn-of-the-century students of inheritance. Among the original membership were plant breeders, animal breeders, seed salesmen, stockmen, university botanists, govern-

[34]There is a sizable literature on biometry, primarily as it developed in Great Britain, but its use by American agriculturalists has not yet been treated. See Daniel J. Kevles, "Genetics in the United States and Great Britain, 1890–1930: A Review with Speculations," *Isis*, 1980, 71:441–455; P. Froggart and N. C. Nevin, "The 'Law of Ancestral Heredity' and the Mendelian-Ancestrian Controversy in England, 1889–1906," *Journal of Medical Genetics*, 1971, 8:1–36; [Student], "The Probable Error of a Mean," *Biometrika*, 1908, 6:1–25.

[35]Kimmelman's "American Breeders' Association" contains an excellent account of the adoption of a Mendelian research program by breeders.

ment researchers, extension workers, and agricultural college deans. And most of them were enthusiastically applying Mendelian laws to plants and animals.[36]

Corn provided a particularly appropriate subject for the early Mendelians. First, unlike wheat or even *Drosophila*, which were both small and hard to manipulate, corn was a large plant with prominent features that were easy to handle and count. Second, its characteristics segregated plainly; with hundreds of kernels to the ear, the breeders had in one plant a large population exhibiting a wide range of variation. The different ways of studying Mendelism in corn were seemingly unlimited. As C. P. Hartley at the Bureau of Plant Industry explained in a 1903 paper, "As corn affords kernels of various colors and of different composition, it furnishes a most interesting subject for the study of hybridization and the transmission of characters."[37]

Early studies of inheritance tended, in fact, to concentrate on those plants and animals that were most appropriate from an experimental point of view, such as rabbits, which reproduced quickly, and corn. R. A. Emerson chose beans "on account of their special fitness for theoretical work." Hartley studied the transmission of characters in corn by breeding for crooked and straight rows of kernels; W. E. Castle demonstrated alternative inheritance in the coat color of rabbits; and H. A. Weber verified Mendel's laws using cotton seeds. Although such studies were conducted more for theoretical than for practical purposes, the methods used were common among the more practically inclined breeders. As one historian has noted, researchers essentially "tallied their F_1s and F_2s and reported consistency with the predicted Mendelian ratios."[38]

[36]Ibid.

[37]C. P. Hartley, "Corn Breeding Work in the USDA," *Proceedings of the American Breeders Association* (*Proc. ABA*), 1903, 1:34. For a fascinating account of the early Mendelian's adoption of corn for more abstract purposes, see Evelyn Fox Keller, *A Feeling for the Organism: The Life and Work of Barbara McClintock* (1983).

[38]Kimmelman, "American Breeders' Association," p. 172; R. A. Emerson, "Bean Breeding," *Proc. ABA*, 1903, 1:53; C. P. Hartley, "Plant Breeding Principles Ap-

But unlike selection and crossing, Mendelism was not, properly speaking, a breeding method in those early years. It was an interpretive tool that breeders could use to inform their selecting or crossing efforts. Thus, while reports from experiment stations were filled with the language of Mendel, what they spoke of were the already-familiar methods.

The stations easily combined their traditional crop improvement interests with Mendelian interpretations. In New Jersey in 1910 B. D. Halsted began studying rules of inheritance in truck crops, focusing especially on tomatoes, while his colleague E. J. Owens studied character transmission in beans, eggplant, and okra. In Massachusetts J. K. Shaw studied garden peas in relation to their heredity, correlation, and variation. J. Belling in Florida attempted to link vigor in beans with the dominance and recessiveness of different characters. In Georgia C. A. McLendon began a cotton-breeding experiment in 1909, in which he observed that "the cotton plant contains some thirty or more pairs of heritable characters, all of which seems [*sic*] to obey Mendel's laws of dominance, segregation, and recombination." During the early 1910s virtually every station working on fruits, vegetables, and grains used Mendelian principles to interpret their breeding results.[39]

While such studies of fruits, vegetables, and field crops seemed to demonstrate the compatibility between traditional breeding methods and Mendelism, corn research told a different story. Beginning with the work of Edward Murray East in 1907 and George Shull in 1908, Mendelism was transformed from an analytical tool into a breeding method. With the introduction of inbreeding and crossing, traditional corn breeders were con-

---

plied to Corn Improvement," *Proc. ABA*, 1905, 2:108–112; W. E. Castle, "Recent Discoveries in Heredity and Their Bearing on Animal Breeding," *Proc. ABA*, 1905, 2:120–126; H. J. Weber, "Explanation of Mendel's Law of Hybrids," *Proc. ABA*, 1905, 2:138–143.

[39]Halsted began this work in 1906. See *Experiment Station Record*, 1910, 22:140–41, 230–31; 1911, 25:436; C. A. McClendon, "Mendelian Inheritance in Cotton Hybrids," *Georgia Exper. Stat. Bull.* 99, 1912, 226.

fronted with a choice of methods. In many other crops selection and crossing continued to coexist and in fact strengthen each other; with corn, however, the alternative ways of studying and improving the crop became competing models by the end of the 1920s.

East and Shull

In many respects, E. M. East was representative of the early twentieth-century breeders who had one foot in the world of practical breeding and the other in the world of inheritance studies. During his years at Illinois, East shifted his intellectual focus from chemistry to plant breeding after working on Hopkins's corn project. Of perhaps greater importance to his later career, however, was the time he spent studying with Charles Hottes. Shortly after East began at Illinois, Hottes, a botanist, returned there after spending a year in Europe studying plant breeding, both directly and indirectly, with Carl Correns, Erich von Tschermak, and Hugo De Vries, while these three were rediscovering Mendel's laws. Under Hottes's tutelege East assimilated the theoretical innovations of the European biologists, and this useful lesson provided the balance wheel to his more practical studies.[40]

When East began work in 1905 at the Connecticut experiment station, his approach to corn breeding was informed by two fairly distinct trends—one more practical and the other more theoretical. Although Hopkins's ear-to-row method of corn improvement was of limited practical use in altering oil and protein content, as a general method of selection and the basis for varietal crossing, it was widely adopted. Meanwhile, during the first decade of the twentieth century, it remained unclear to breeders how the theoretical work of Mendel, De Vries, and other experimentalists

[40]Cannon, *Great Men of Modern Agriculture*, p. 221; Crabb, *Hybrid Corn Makers*, p. 23.

could be put to practical use in corn improvement. The agenda for East, then, was to bridge the gap between the two by infusing practical corn breeding with a Mendelian spirit.

Between 1906 and 1908 in the Connecticut station bulletins and reports East published five papers that aimed to do just that. The first, written during his first year there, was a discussion of the Illinois corn project and its application for Connecticut corn growers. The aim of the breeder, according to East, was "to provide the best methods for recognizing, selecting, and propagating such favorable variations as occur" in the cornfield.[41] East believed the best method combined Hopkins's ear-to-row and Beal's varietal cross; the first allowed the breeder to control both male and female parents and to produce a repeatable pedigree, and the second increased favorable characteristics. In selecting superior seed for crossing, East recommended that the initial selection be made in the field rather than from the corncrib because the breeder needed to consider not just the kernel and ear appearance but the total plant. Good breeding corn was characterized by a strong, well-rooted cornstalk; broad foliage; ears at medium height on the stalk; deep, uniform, and highly viable kernels. Referring to the show-card method of selecting corn, East declared that there was positively no correlation between yield and kernel appearance.[42]

Turning to the breeding itself, East identified three separate methods: inbreeding, close (or sibling) breeding, and crossbreeding, by which he meant varietal crossing. He warned against "the evil effects of close and inbreeding," suggesting that corn growers perform varietal crosses by detasseling alternate rows.[43] He also recommended that growers judge the corn by the average weight of the ears in each row, in order to take into account barren and diseased as well as healthy corn. Since each row corresponded to

[41]E. M. East, "The Improvement of Corn in Connecticut," *Connecticut AES Bulletin* 152, 1906, 6.

[42]Ibid., pp. 9–11.

[43]Ibid., p. 14.

a particular ear, the breeder could fairly accurately determine which ears were superior and which were not.

This was an innovative method of selecting corn; most contemporary breeders considered only the physical appearance of individual ears, ignoring both plant appearance and progeny, or yield, and thus considerably slowing improvement. One ear of corn usually carried both positive and negative characteristics, not all of which were apparent on physical examination. The ear-to-row method permitted the breeder not only to identify the parents but also to see the range of variability within any given ear. An ear-to-row test could demonstrate, for example, that most of the kernels on a promising-looking ear in fact produced diseased plants or weak root structure or barren stalks—all of which would lower the row yield. East's method infused a new level of accountability in corn testing.

As a demonstration of the negative effects of inbreeding, East recommended comparing the row yield from both tasseled (inbred) and detasseled (crossbred) rows. This test showed that "crossed rows yielded in the second and third years an average of ten bushels per acre more than tasseled rows."[44] Few breeders would have been surprised that inbred rows yielded poorly, but East's claim for a ten-bushel increase from varietal crossing confirmed Beal's earlier work and must have inspired some growers to alter their methods accordingly. East's paper then concluded with elaborate instructions on how to plant breeding plots, increase plots, and commercial fields. The increase plot consisted of seed that had proven superior in the breeding plot, which was then planted en masse to increase the quantity of seed. The commercial field consisted of corn from the increase plot, which had been selected for a third year and could be sold as registered, pedigreed seed.[45]

East's plan for corn improvement was, from a grower's point of view, exceedingly complex. In an effort to prevent inbreeding and

[44]Ibid., p. 12.
[45]Ibid., p. 18.

control varietal crossing, East provided a field rotation chart that required growers to keep painfully exact records of which row numbers corresponded with which ear numbers and then to systematically change the rows the following year. For example, field row number 1 would be planted with seed from last year's row number 76 one year, and from row 78 the following year; row number 51 would be planted from seed grown in row number 2 one year, and row number 4 the following, and so on. In addition, growers were supposed to divide the fields into quarters and rotate the tasseled (inbred) rows into another quarter each year. All in all, it seems unlikely that any but the most compulsively progressive farmers would have had the patience or the wit to see the plan through.

Having dispensed with practical corn growing, the following year East turned his attention to more theoretical considerations. Reviewing the previous ten years of plant breeding and biology, East noted the lack of correlation between breeders and theorists: "Theorists were too busy in their debates to obtain many results of practical value to the plant breeders; and the plant breeders were obtaining so many surprising complications in their experimental work that the thought of the possibility of classifying their thousands of observations under a few natural laws did not occur to them."[46] Those puzzling over this lack of unified effort, however, needed only to follow East's tour of the literature to learn why the two groups were apart. Beginning with Lamarck and Darwin, East ticked off the contributions to biological thought made by Weismann, Malthus, De Vries, Wilhelm Johannsen, Mendel, Galton, T. H. Morgan, Bateson, and Hopkins, certainly a bewildering collection of thinkers, whose aggregate work hardly cleared the muddy waters. East, however, thought the lessons to be learned from them were nothing short of revolutionary. Referring especially to Mendel, he claimed that "experimental biology has had a door of knowledge opened comparable to

[46]East, "The Relation of Certain Biological Principles to Plant Breeding," *Connecticut AES Bulletin* 158, 1907, 5.

that which Dalton opened for chemistry with his atomic theory."[47]

From the many approaches to and explanations of heredity, East culled three principles he considered basic for plant breeders interested in the scientific improvement of cereals and grains. First, patterns of inheritance could be interpreted in terms of the transmission of unit characters in a Mendelian fashion. Second, breeding does not create new characters but rather reshuffles those already in existence. And third, the selection of such "fluctuations" can increase or decrease the frequency with which such characters appear in the plant.

In a sense East was extracting those components of Mendelism and biometry most useful to breeders and combining them in a way that was particularly critical for agriculturalists. Despite the appearance of contradiction in that "Galton's law is statistical and deals with averages [while] Mendel's law is physiological and deals with individuals," agricultural breeders needed the contributions of both to generate commercially significant improvements.[48] The breeder trying to select for a particular characteristic needed to focus first on the individual exhibiting the characteristic and could use Mendel's laws of dominance and probability in bringing such a character into prominence. Once that character was isolated, biometric techniques of prediction within a population could guide efforts to produce an entire population exhibiting the characteristic.

In a series of three papers the following year East attempted to bring theory and practice into alignment. "The time has come," he announced, "when the seed corn producers of the United States, who are to be successful in their work, will be the men who study and apply the underlying principles of variation and heredity, and among these laws Mendel's law of dominance is worthy of consideration. Its use in corn breeding is very simple, and its study should prevent many fallacious proceedings and

[47]Ibid., p. 64.
[48]Ibid., p. 44.

render the work of corn improvement more methodical."[49] And indeed, Mendel's law of dominance was simple to apply to corn; East had discovered that flint corn was dominant to both sweet and dent corn, that dent was dominant to sweet, and that purple kernels were dominant to white. He found, moreover, that a recessive character, such as white kernels, could be isolated by inbreeding. When the entire ear exhibited white kernels, it was assuredly pure and by continued careful breeding would stay that way.

With the aid of a Mendelian framework, East was beginning in 1907 to reevaluate the role of inbreeding in plant improvement. Before the Mendelian interpretation, East as well as his contemporaries accepted Darwin's strictures against inbreeding as self-evident; East's own experiments showed clearly that inbreds were weak, stunted, and misshapen, whereas varietal crosses were vigorous. Mendel's work suggested a different use for inbreeding: if one could consider plant characteristics as unit characters variably existing within the population and if such a character could be selected for, isolated, and its rate of frequency within the population increased by inbreeding, then inbreeding was a powerful tool in purifying strains and reducing them to the desired unit characters.

East was not the only scientist who was rethinking inbreeding as a Mendelian tool. At Cold Spring Harbor, New York, George Shull was intently studying the same problem. Shull, a plant physiologist who had received his Ph.D at the University of Chicago in 1904, was assisting Charles B. Davenport at the Station for Experimental Evolution. Both Shull and Davenport were keen students of Mendelism, and one of Shull's first assignments was to breed corn that could be used to demonstrate Mendel's principle of segregation. Unlike East's, Shull's use of corn did not spring

[49]East, "Report of the Agronomist: II. Practical Use of Mendelism in Corn Breeding," *Connecticut AES Report, 1907–1908*, p. 418. See also his "Report of the Agronomist: I. The Prospects of Better Seed Corn in Connecticut," ibid., pp. 397–405, and "Report of the Agronomist: III. Inbreeding in Corn," ibid., pp. 419–428.

from an interest in breeding and practical improvement; indeed, the reverse was true. Shull's main interest was the principles of inheritance, and his choice of corn grew out of both Davenport's preferences and the apparent utility of corn as an experimental subject. But like East, Shull went on to combine theoretical and practical interests as it became clear that corn could be substantially altered through a Mendelian approach.[50]

In 1904 Shull began studies of the number of rows of kernels per ear, in which he compared inheritance of row number in inbred and crossbred corn. He used the ear-to-row method, planting a row of ten-row-ear inbred corn next to a ten-row-ear varietal cross, both of which had "as nearly equivalent parentage" as possible. The next two rows would have a twelve-row inbred and a twelve-row cross and so forth. When the plants matured, he compared the inbreds to the related cross and to each other according to their manifestations of several characteristics. He observed that the inbreds not only differed considerably from one another in these characteristics but differed from the related variety. Shull reasoned that "if self-fertilization is assumed to be the direct cause of any of the above characteristics of the one inbred row, it is obviously illogical to attribute the opposite characteristic possessed by the other row to self-fertilization as a direct result." In other words, since the progeny of inbreds exhibited characteristics that were not apparent in the original variety, self-fertilization could not be considered a mechanism for concentrating inferior characteristics. His interpretation coincided with Johannsen's pure-line theory, according to which inbreeding simply isolates biotypes that exist within a mixed strain. Shull concluded that "an ordinary cornfield is a series of very complex hybrids produced by the combination of numerous elementary

[50]Shull's corn-breeding project has been briefly sketched in several places. See Henry A. Wallace and William L. Brown, *Corn and Its Early Fathers* (1956), pp. 106–111; Paul Manglesdorf, *Corn: Its Origin, Evolution, and Improvement* (1974), p. 212; Sprague, *Corn Improvement*, p. 233; and Crabb, *Hybrid Corn Makers*, pp. 45–48.

species. Self-fertilization soon eliminates the hybrid elements and reduces the strain to its elementary components."[51]

At the 1908 meeting of the American Breeders' Association, Shull presented his findings and suggested the implications for breeders. Although he had apparently not crossed the inbreds he had produced, it was clear to him that such a procedure could dramatically improve corn strains then in use by breeders. His suggestion for further research reflected a familiarity with contemporary breeding techniques as well as a confidence in the utility of his own work for breeders: "The fundamental defect in every empirical scheme of corn breeding which simulates the isolation [ear-to-row] method of the breeder of small grains, lies in the fact that there is no intelligent attempt in these methods to determine the relative value of the several biotypes *in hybrid combination* but only in the pure state."[52]

One of those in attendance at Shull's presentation was East, who, following the conference, borrowed the paper from Shull so he could "study your results before spring planting."[53] Not coincidentally, East was also preparing the third part of his report for the Connecticut station, titled "Inbreeding in Corn," in which he would again review the literature, but this time with the benefit of Shull's work on inbreeding. East was somewhat chagrined at Shull's clever interpretation of experiments that closely resembled East's own, and "wonder[ed] why I have been so stupid as not to see the fact myself." Shull, for his part, made a poor attempt at graciousness when he agreed that East's experiments "might have led . . . to the same conclusion" and then, casting grace—and modesty—aside, claimed that "if I had held on to my idea of the composition of a field of corn until I could have worked out some

[51]George Shull, "The Composition of a Field of Maize," *Proc. ABA*, 1908, 4:298, 299.

[52]Ibid., p. 300.

[53]East to Shull, 5 February 1908. All East-Shull correspondence referred to here is reprinted in Donald F. Jones, "Biographical Memoir of Edward Murray East," *National Academy of Sciences Biographical Memoirs*, 1944, 23:217–242.

of the subsidiary problems, I could have raised a monument to myself which would be worthy to stand with the best biological work of recent times. But the matter seemed to me of too great importance in view of the value of our maize crop to selfishly keep it to myself longer." East was encouraged, nonetheless, by Shull's work, and in his paper he credited Shull with giving him the biotype interpretation of inbreeding.[54]

East and Shull continued to conduct corn-breeding experiments and occasionally compared notes. In June 1908 East visited Shull at Cold Spring Harbor, and in 1909 Shull went to Connecticut with von Tschermak and K. von Rumker, a German plant breeder, to investigate East's experiments. But their relationship became increasingly strained. Both were publishing papers at a rapid pace, and the question of priority was a frequent source of irritation on both sides.[55]

In the debate over whether East or Shull deserves credit for launching hybrid corn studies, historians have lost sight of the more important issue of the differences in their agendas and how these agendas guided their research. East, though he applauded Shull's idea of crossing inbreds, felt that its value was theoretical rather than practical. To East, whose primary commitment was to the experiment station and its clientele, hybrid seed seemed an attractive but unlikely solution to the corn improvement problem. For one thing, it relied on a complicated method that would discourage breeders. East at times seemed exasperated that Shull, whose only commitment was to his scientific peers, failed to grasp the practical constraints: "I wish you could have a little experience trying to get the farmers to take up anything in the least complex, and I know you would agree with me that only the very simplest things can be done by the corn grower." The method, moreover, was costly as well as complex. Inbred ears

[54]East to Shull, 12 February 1908, Shull to East, 3 March 1908.

[55]Jones, "Biographical Memoir," p. 224; Herbert Kendell Hayes, *Professor's Story of Hybrid Corn* (1963), p. 9.

were so dwarfed that their yield was correspondingly small, and growers could not possibly receive a fair price for such seed. Until these limitations could be overcome, East felt, growers would be well advised to produce varietal crosses rather than hybrids.[56]

Shull and East continued their corn studies for several more years, Shull championing the pure-line method of breeding and East recommending varietal crosses. Shull published his last substantive paper on corn in 1911 and then returned to his first experimental love, the primrose. His four-year study of corn inheritance had contributed greatly not only to programs in theoretical genetics but to practical studies of corn as well, although it was not until 1918 that his pure-line method of improvement found practical application.[57]

For Shull, as for many other biological scientists, corn was simply a means of exploring Mendelian phenomena, just as *Drosophila* had been for Thomas H. Morgan. Because Shull was interested not in corn improvement per se but in hereditary laws, he was able to move easily from one experimental subject to another, uncovering in each its particular patterns and mechanisms of inheritance. This is not to suggest that corn was a trivial part of Shull's overall research program. His development of the concept of heterosis, or hybrid vigor, which was his most important—and controversial—contribution to genetics, was based on his early work in corn breeding. But for Shull corn was the window, not the landscape.[58]

[56]East to Shull, 4 February 1909. See also East, "The Distinction between Development and Heredity in Breeding," *American Naturalist*, 1909, 43:173–181.

[57]Shull, "The Genotypes of Maize," *American Naturalist*, 1911, 45:234–252; "Experiments with Maize," *Botanical Gazette*, 1911, 52:480–485.

[58]The heterosis concept is treated at length in J. W. Gowan, ed., *Heterosis* (1952). Jean-Pierre Berlan and Richard Lewontin have been highly critical of Shull and the heterosis idea; see, for example, Berlan, "Heterosis and Hybrid Corn: Much Ado about Nothing?" (Manuscript, March 1986); Berlan and Lewontin, "Political Economy of Hybrid Corn."

East, too, was ultimately more interested in heredity than in corn, but his legacy was somewhat different from Shull's. In 1909 he was appointed professor of plant morphology at Harvard's Bussey Institution, where his interest shifted to inheritance in the tobacco plant. He continued to direct experimental studies at Connecticut until about 1915, when his last corn paper appeared. More important, he took on two graduate students: one was to provide the long-sought-after key to commercial hybrids, and the other was to devote his professional career to aligning corn genetics and farming practice.

Herbert Kendell Hayes began working with East in the summer of 1909, hired by the station to oversee the experiments East had designed while at Harvard. Hayes, who had been working on a tobacco-breeding project for the USDA in northern Connecticut, learned the principles of breeding and heredity both in the classroom at Harvard and in the cornfields of Connecticut. The last four papers East produced (three for the Connecticut station and one for the Bureau of Plant Industry) were jointly authored with Hayes and continued East's earlier interest in applying Mendelian formulations to practical breeding concerns. Their commitment to practice, reinforced and perhaps explained by their position at an experiment station, was evident in nearly all their reports. In 1911 East and Hayes appealed directly to breeders. Noting that "maize is the basis of the agricultural wealth of the country," they declared it "eminently desirable that every fact about the inheritance of its characters should be learned as soon as possible. It is only through the application of such knowledge that the present arbitrary, and, in a way, unscientific, methods of its improvement as an economic crop will be placed upon a definite and orderly basis."[59]

East and Hayes saw that the problems of inheritance faced by the corn breeder had a much broader significance in agriculture

[59]E. M. East and H. K. Hayes, "Inheritance in Maize," *Connecticut AES Bulletin* 167, 1911, 9.

and that the questions they asked might be posed by breeders and biologists alike: do all characters in corn observe the same hereditary laws? are some characters correlated or linked with others in inheritance? which characters are dominant and which recessive, and at what point, if ever, can such characters be considered pure? For at least another decade such questions formed the core of Mendelian studies regardless of the organism studied or the degree of practical significance involved.

East and Hayes continued to push their idea of varietal crossing for commercial purposes, although without apparent success. Admitting that it "has been suggested time and again without gaining a foothold," they nonetheless felt it offered more immediate gratification than either selection or Shull's single cross. They continued inbreeding corn, although largely for the purpose of studying segregation and dominance, and in 1912 discovered that aleurone color was modified by at least four other factors.[60]

Inbreeding provided a potent tool for isolating such characteristics, and East and Hayes used it extensively to map hereditary patterns, most of which lacked economic value. Practical breeders and farmers, who had been trained to avoid inbreeding at all costs, must have looked askance at experiments that appeared to make corn worse rather than better. In 1912 East and Hayes were moved to defend inbreeding as a purifying agent: "Poor naturally cross-fertilized strains survive the scythe of natural selection; they are selected for propagation by man because they are crossed with other strains and are vigorous through heterozygosity. Inbreeding tears aside their mask. They must then stand or fall on their own merits. . . . The poor strains that have had the help of hybridization with good strains, combined with the added vigor due to heterozygosity, are stripped of all pretense, shown in all their nakedness, and inbreeding is given as the cause of the degeneracy."[61]

[60]E. M. East and H. K. Hayes, "Heterozygosis in Evolution and Plant Breeding," *USDA Bureau of Plant Industry Bulletin* 243, 1912, 47.

[61]Ibid., p. 38.

The Business of Breeding

The Conjunction of Theory and Practice

By around 1910, corn studies combined the research techniques and interests of two different kinds of breeders. The first was the practical breeder or the farmer, who, often without the benefit of a scientific education, attempted to improve corn yield by means of selection and varietal crossing. Farmers, seed-corn breeders, and seed sellers had a purely practical interest in corn. The second kind of breeder was the botanist or biologist, generally affiliated with an agricultural college, who used corn as a scientific device to study the general laws of inheritance. These breeders may have applied their academic skills to more practical concerns, but their long-term interest was scientific rather than agricultural. The scientists whose work culminated in hybrid corn dwelled in a peculiar juncture of pure and applied science, an institutional juncture, for East, Hayes, and later Donald Jones were agricultural experiment station scientists. By training and affiliation, their commitment was to the scientific solution of practical problems.

In 1910 none of the corn improvement methods then under consideration—selection, varietal crossing, or single-cross hybrids—could claim an ability to increase the corn yield dramatically. Indeed, none were designed to do so. Only selection had been practiced long enough to report reliable, if gradual, improvement in the corn crop. Breeders were ambivalent about the potential for varietal crossing, though it was created to improve yields. Single-cross hybrids were invented as much to satisfy experimental curiosity as to fulfill farming needs, and in 1910 the experimental evidence in favor of hybrids was slim indeed.

2

The Meaning of the Method:

Corn Improvement at the Bureau of Plant Industry

To say that corn was important to the United States would be a gross understatement; corn during this period was the most economically important crop in American agriculture. "The value of this crop," wrote the secretary of agriculture in 1908, "almost surpasses belief. It is $1,615,000,000. This wealth that has grown out of the soil in four months of rain and sunshine, and some drought, too, is enough to cancel the interest-bearing debt of the United States and to pay for the Panama Canal and fifty battle ships."[1] Within the USDA Bureau of Plant Industry, as at the agricultural experiment stations in the corn-growing states, it was studied from every conceivable angle. And though it would be a mistake to assume that experiment stations eagerly adopted the research ideas defined by the USDA, the USDA did foster trends and ideas that might have never survived without its authority. In corn work the USDA spoke with the same progressivistic fervor it used for nearly every other agricultural issue. The message to experiment stations, legislators, and farmers was an unwavering commitment to abundance, scientific advance, efficiency, and prosperity.

Yet if the rhetoric of the USDA was forward-looking, its activities—not surprisingly, considering its role as the first authority on agricultural questions in the United States—were more con-

[1]James Wilson, "Report of the Secretary," *Yearbook of Agriculture*, 1908, 10.

servative. Nevertheless, the Bureau of Plant Industry set out to explore the new Mendelian approach while maintaining a more traditional corn improvement program. Just as the state experiment stations continued their selection experiments and field trials while branching out to study the Mendelian theories of segregation and dominance, so the BPI initially added Mendelian studies to the existing program in traditional crop improvement.

But during the period from 1909 to 1919 the dynamic between these two approaches underwent a profound transformation. By the end of the decade, the Mendelian approach, which had seemed aimlessly experimental, was the research method of choice; conversely, the popular corn improvement work faltered throughout the decade until it was essentially killed off in the early 1920s. What began as a set of complementary ideas about how the corn crop might be improved ended up as alternative and competing methodologies. Although the personalities involved in this transformation might appear more influential than the ideas they represented, there can be no doubt that Mendelism in particular represented more than scientific method to the BPI. Indeed, it promised to revolutionize not only the yield of corn but the very process of producing it.

C. P. Hartley and the Persistence of Tradition

As physiologist in charge of corn investigations at the BPI from 1906, Charles Pinckney Hartley set the tempo for federal corn research in the early 1900s. Born in the rural Midwest, Hartley had received a master's degree from Kansas State University before joining the BPI as assistant physiologist in 1899. He was an early member of the American Breeders' Association, though ambivalent about Mendel's theories. The possible advantages of Mendelism for practical breeders were not at all self-evident as far as Hartley could see, and he was cautious in discussing how Mendelian studies might contribute to traditional corn improvement practice.[2]

In a 1909 article Hartley's dissatisfaction with Mendelian theory was thinly disguised behind a poorly reasoned assessment of Mendelism's shortcomings. Though he maintained that selection and Mendelian theory were perfectly compatible, Hartley nonetheless felt sure that selection would always be the critical element in breeding because even Mendelians must start with good corn strains. Maybe Mendelian theory is right, maybe it is not, Hartley hinted, but selection will always be with us. Hartley cautioned against the dangers of elegant abstract principles that might blind breeders to the more mundane but inescapable problems in crop improvement. "It is highly important for plant breeders to know that certain characters obey Mendel's law and to be able to distinguish the dominant characters from the recessive, and it is necessary to bear in mind that under differing conditions the degrees of dominance or recessiveness may differ or cease to exist."[3]

Hartley's dubious logic reflected not only his personal distaste for the new Mendelian theory but also his concerns for the organizational safety of his own corn improvement program. His style of federal science was somewhat unusual even for this period, and closely resembled the style apparent at many agricultural experiment stations. Educated in the years before Mendelism, Hartley was an old-fashioned corn breeder who emphasized cooperative field trials and personal attention to participating farmers in the states. He directly supervised field experiments in many states, gave farmers advice, and lectured on the virtues of self-help and hard work. Hartley seems to have had no experience with laboratory science and not much interaction with the more scientifically focused community of breeders. In short, his

[2]"Charles Pinckney Hartley," *American Men of Science*, 6th ed., pp. 606–607.

[3]C. P. Hartley, "Progress in Methods of Producing Higher-Yielding Strains of Corn," *Yearbook of Agriculture*, 1909, 315.

attention was given to practical corn breeding rather than principles of breeding. It was a distinction viewed in some quarters as a conflict between old-fashioned and modern corn breeding, between amateur and scientist.

Having disposed of the pesky problem of somehow unifying Mendelian theory and crop improvement practice, Hartley was free to continue his efforts to increase corn yield directly. In another article, the following year, he identified the two most important objectives of breeding experimentation. First, farmers should simply "compare the productive power of different ears of corn"; this exercise required very little expert knowledge and only a little extra attentiveness at harvest. Second, Hartley provided directions for breeders wanting to crossbreed those varieties or lines that seemed to yield especially well. Together, these efforts would select and maintain, perhaps even increase, high-yielding corn and at the same time would reduce the amount of poor corn in the field. The procedure couldn't have been less theoretical.[4]

Hartley's network of cooperating farmers was scattered throughout the states, but he concentrated on farmers in Ohio. In the early 1910s Hartley traveled to Ohio twice a year "for the purpose of taking notes, directing the detasseling, and performing hand-pollinating work in connection with corn experiments."[5] He visited other states as well at planting and harvest time to help in seed selection and to see how USDA seed had fared in different geographical regions. In his occasional addresses to farmers, such as that to the Butler County, Ohio, Corn Improvement Association in the winter of 1912, Hartley hinted broadly at his belief in the natural relationship among corn research, corn growing, and moral virtue:

[4]C. P. Hartley, "Directions to Cooperative Corn Breeders," *Bureau of Plant Industry* 564, *Corn Investigations*, 23 April 1910.

[5]C. P. Hartley, "Letter of Instruction," 8 July 1911, record group 54, entry 2, box 518, National Archives (hereafter 54/2/518, NA). Hartley's travels were noted in official authorizations found in box 319.

> It has been said to me that our work benefits grain speculators but does not benefit farmers, because better yields simply mean lower prices. Social and commercial situations influence prices but can trouble the farmer only as he permits them. Being in the majority and having first possession of the necessities of life, failure to acquire as much knowledge and exercise as much intelligence as his rivals can alone prevent him from being master of these situations. He can master these situations by striving eight hours a day for wisdom rather than sixteen hours a day for gold.[6]

If Hartley sounded rather like a missionary trying to eradicate whatever residual paganism his flock might be harboring, it was an apt reflection of his embattled stance. By pitting virtue, thriftiness, and stability against greed and reckless ambition Hartley was responding not only to the farmers' traditional fear of speculators but to his own fear of scientific gee-whizzery and the instability it threatened. His message, though somewhat confusing, was not really an assurance that corn prices would not fall with increased yields but rather a warning that farmers should not emulate grain speculators by trying to increase yields inordinately.

Hartley's concerns for the future of his crop improvement program were not unfounded. In the fall of 1912 Bureau Chief Beverly T. Galloway took Hartley to task for spending so much time on local problems, especially those in Ohio. Galloway argued that Hartley should focus more attention on "broad general principles" rather than particular corn varieties, suggesting that "the only project which would seem to be of a general nature is the one having to do with the Effects of Heredity and Environment on corn, which, so far as the project sheets show has—probably for financial reasons—been forced to take a somewhat subordinate

[6]C. P. Hartley, "How Can the USDA Best Help the Corn Growers of the United States" (address to the Butler County, Ohio, Corn Improvement Association, delivered 10 or 11 January 1912, 54/2/319, NA).

place."[7] Hartley was happy to agree with Galloway that the real problem was not intent but financial need, responding that "financial reasons alone compel us to give much of our own energies to the accomplishment of immediately profitable results somewhat to the neglect of investigating fundamental principles—our proper and preferred field of work." Several months later Hartley again defended his program, claiming that it was concerned not with "the active production of a good corn crop, but with the networks of its production and the influencing causes."[8]

By pitching his working definition in terms of methods and causes, Hartley was clearly trying to convince Galloway that the corn program was more theoretically grounded than, in fact, it was. Meanwhile, he did not redirect his efforts to bring them more into line with Galloway's suggestions (the phrase "Effects of Heredity" in contemporary parlance could mean only Mendelian studies). Hartley continued to champion selection work, claiming that his office had helped Ohio farmers increase yield by ten to twenty bushels an acre since 1900 and that several hundred acres in Ohio would yield over a hundred bushels an acre in 1912 because of his efforts. Even the secretary of agriculture's annual report contained the astonishing claim that selection alone had increased the yield of some varieties by 25 to 30 percent, a figure that no one but Hartley could have—or would have—provided.[9]

At the end of 1912 Hartley filed his report on research results with Galloway; the document was as much a report on the past as an agenda for the future. Hartley emphasized that the ear-to-row method of selection and breeding was most promising and that varietal crossing had real potential. Avoiding the question of crossing inbreds, which East and Shull had discussed four and five years earlier, Hartley merely declared that inbreeding would

[7]Beverly T. Galloway to Hartley, 17 October 1912, ibid.

[8]Hartley to Galloway, 26 October, 9 December 1912, ibid.

[9]Hartley, Address to Butler County, Ohio, Corn Improvement Association, Hartley to Galloway, 14 October 1912, ibid.; James Wilson, "Report of the Secretary," *Yearbook of Agriculture*, 1912, 128.

produce disastrous results. And so Hartley's work continued. He filed periodic reports on varietal crossing and selection and boasted about one USDA varietal cross that was widely adaptable and almost uniformly better than local varieties. Hartley also prepared a bulletin aimed at young boys, titled "How to Grow an Acre of Corn," in which he laid out the same principles of selection he provided for their fathers.[10]

By the mid-1910s, however, Hartley's was increasingly a minority viewpoint within the USDA. As early as 1904, G. N. Collins was avidly studying corn from a Mendelian viewpoint. In a study begun in 1909, he found a pure white ear of corn in a field of selected yellow corn. This phenomenon, which he identified as "incomplete segregation," was "an obstacle to securing combinations of characters by hybridizing," and he suggested that selection might still be necessary "to establish the full expression of characters."[11] With J. H. Kempton, Collins studied the transmission and segregation of characters in a cross between American and Chinese corn, and considered how crossbreeding affected the size of corn seed. Although none of these projects could reasonably be interpreted as threats to Hartley's program, neither did they seem to reinforce the idea that corn improvement could progress with less theory and more field trials.[12]

Anybody Can Cross Corn

While Hartley may have felt rather defensive about the power of his research style compared to that of the new Men-

[10]Hartley to Galloway, 9 December 1912, 54/2/319, NA; Hartley, *Bureau of Plant Industry Bulletin* 218, 72; Hartley, "How to Grow an Acre of Corn," *USDA Farmers' Bulletin* 537, 1913; Hartley to W. A. Taylor, 23 December 1913, 15 September 1914, 54/2/319, NA.

[11]G. N. Collins, *Bureau of Plant Industry Bulletin* 272, 1913; G. N. Collins, "Increased Yields of Corn from Hybrid Seed," *Yearbook of Agriculture*, 1910, 319–328.

[12]G. N. Collins and J. H. Kempton, *Bureau of Plant Industry Circular* 124, 1913, 9–15.

delians in the USDA, he also may have derived some comfort from the consonance of his approach with that of agricultural experts in the corn-growing states. In the mid-1910s, Mendel was a stranger to farmers, and the value of Mendelian theory to practical crop improvement seemed fairly dubious even to those breeders who were familiar with it. In the pages of *Wallaces' Farmer*, for example, the discussion of corn improvement turned on more traditional issues such as selection standards and crossing varieties. What is most striking about these reports is the persistent idea that farmers themselves were participants in the process of crop improvement. Hartley's belief in the value of individualistic, voluntary corn improvement by farmers in their own fields with the encouragement and advice of the USDA was widely shared by contemporary breeders, most notably Henry A. Wallace.

Wallace's interest in corn was piqued by Perry Holden, the self-styled "corn evangelist" who came to Iowa State University from the University of Illinois at the turn of the century. When Wallace was sixteen, Holden persuaded him to try some experiments in corn breeding, which Wallace found fascinating and perplexing. He entered Iowa State two years later to give the subject more rigorous treatment under the guidance of H. D. Hughes and Martin Mosher. Wallace was graduated in 1910, traveled through Europe in 1912, and in 1913 began considering the role of Mendelian theory in corn improvement with his first experiments using inbreeding.[13]

One of the reasons Wallace provides such a useful and interesting view onto the world of corn improvement is that, during this period at least, he seems to have considered all approaches to breeding and improvement equally valid. For Wallace, corn improvement represented both a practical and a theoretical challenge—practical because he was concerned with educating farmers to treat their corn crop with more intelligence and care, and

[13]Edward L. Schapsmeier and Frederick H. Schapsmeier, *H. A. Wallace of Iowa: The Agrarian Years, 1910–1940* (1968).

theoretical because, as a college-trained breeder, he was fascinated with the underlying logic of inheritance in corn. It was the practical approach, of course, that appeared regularly in *Wallaces' Farmer* and earned him the admiration and respect of most farmers in the Midwest.

Wallace's belief in corn improvement and the role of farmers was epitomized in a letter to the editor of *Wallaces'* in early 1915. Peder Pederson, an Iowa farmer,[14] laid out for *Wallaces'* readers not only the opportunities but the responsibilities of farmers to become involved in corn improvement. It is worth quoting at length, I think, because it aptly summarizes the attitude of "progressive" farmers and agricultural leaders alike. After describing his own varietal test plots, Pederson wrote:

> In our opinion, it is too bad that so few farmers will avail themselves of the splendid opportunities to try out small grains and corn for themselves in their own farms, under the direction and with the aid of the extension department at Ames, and the Iowa Corn and Small Grain Growers' Association. If that work was taken up as a rule, and not as an exception, by us Iowa farmers, we could and would in a few years vastly increase our yields of small grain and corn. . . . the only way for us farmers to get any good from the magnificent work done by the extension department is for us in a small way, each on his individual farm, to take hold of and try for himself the varieties tried and recommended from Ames. . . . Brother farmers, let us all get down to make at least one trial each year, either with corn or small grain; and let us faithfully report the results, be they favorable or unfavorable, then we shall soon know, each and every one of us, what kinds and varieties are the best suited for our locality.[15]

In Iowa and elsewhere, farmers were constantly exhorted to use such field trials, as much to provide the experiment station with

[14]Although Pederson was identified as an Iowa farmer, the tone of his letter suggests that he was invented by the Iowa State Agricultural Experiment Station.

[15]Peder Pederson, letter to the editor, *Wallaces' Farmer*, 26 February 1915, 369.

data as to improve their own favorite corn lines. With the many different field and environmental conditions across each state and the many different strains of corn in use, most stations were simply incapable of learning which lines produced well and which poorly in any given locality. The hope was that farmers' self-interest and curiosity would create a corps of untrained but highly knowledgeable farmer-cooperators who would help the stations assemble facts and figures on corn productivity patterns across the states.

Henry Wallace was an enthusiastic supporter of such experimentation and repeatedly urged farmers to participate in what must have seemed to some a campaign. Although Wallace was also enthusiastic, or at least opinionated, on virtually every other aspect of farming as well, his most enduring attention was to corn. Thus it is not surprising to find in nearly every issue of *Wallaces' Farmer* some discussion of corn growing. His response to a farmer's question about varietal crossing in 1915 was typical. Pointing out that "most of our best varieties of corn are the result, not of crossing, but of selecting," he nonetheless went on to suggest that some farmers might want to experiment with varietal crossing because "our theory is that it may possibly pay the farmer on account of the increased yield." The following year, however, when Wallace again gave directions for crossing varieties, he cautioned farmers that there might easily be no improvement in yield.[16]

As at the experiment stations and the USDA, in *Wallaces' Farmer* ear-to-row selection and varietal crossing were the subjects of much debate. Wallace couldn't seem to decide whether they were worth the effort or not from an economic point of view, but he generally recommended that farmers try them out for fun anyway. "It is intensely interesting to experiment with the cross-breeding of seed corn," Wallace wrote in 1916, "and not so very

[16]"Corn Breeding," *Wallaces' Farmer*, 26 February 1915, 379; "Crossing Corn," *Wallaces' Farmer*, 12 January 1916, 1.

difficult." As he reassured readers of "The Boys Corner," a special column for farm boys, "anybody can cross corn."[17]

The one crop improvement strategy Wallace did not waver about was the selection of corn not by the corn show scorecard but by productivity of ears. Wallace reported at length on the results of an Ohio experiment that explored the correlation between show corn and high yield and was gleeful that the Ohio station verified his own dissatisfaction with show standards: "These experiments upset a lot of cherished convictions in the minds of seed corn breeders. Many farmers, however, after reading the account . . . will say: 'I told you so.'" At the BPI, Hartley had also warned his cooperating breeders against selecting for scorecard standards, and in Illinois Eugene Funk voiced the same sentiment. It must have been a hard lesson to teach, nonetheless; as late as 1917 Wallace again urged farm boys to grow corn for yield rather than appearance, appealing to their desire to be superior to the other boys: ordinary boys could grow for corn show appearance, he suggested, but clever boys would grow for yield.[18]

As time went on, however, Wallace began to lose his appreciation for both varietal crossing and ear-to-row selection as corn improvement methods. Despite encouraging reports from some farmers, which were printed in *Wallaces' Farmer*, Wallace himself was concluding that "ear-to-row work is not the short and certain road to improvement that some people would have us think."[19] Two years later, Wallace was even more convinced:

[17]"Cross Breeding Corn," *Wallaces' Farmer*, 21 January 1916, 94; "Corn Breeding Plot," *Wallaces' Farmer*, 29 March 1918, 578.

[18]"Corn Experiments," *Wallaces' Farmer*, 30 April 1915, 700; "Telling the Yield of Corn by Its Looks," *Wallaces' Farmer*, 22 December 1916, 1676; C. P. Hartley, "Directions to Cooperative Corn Breeders," *Bureau of Plant Industry* 564, 23 April 1910; Funk's views were reported by Wallace in "Show Corn and Yield," *Wallaces' Farmer*, 13 April 1917, 656–657; "The Corn Breeding Plot," *Wallaces' Farmer*, 12 January 1917, 58.

[19]See, for example, George M. Allee, "Some 1916 Corn Experiments," *Wallaces' Farmer*, 23 February 1917, 351–352; "Some Interesting Corn Hybrids," *Wallaces' Farmer*, 7 March 1919, 587, 590, in which "hybrids" refers to ear-to-row corn,

"Hundreds of men in the corn belt have conducted ear-to-row tests in an effort to select their corn for high yields. For a time, most of these men have been enthusiasts, but after a few years they have discovered that corn bred for yield is little if any better than corn which has not been selected especially for yield." Varietal crossing seemed no better, and Wallace cited L. H. Smith at the University of Illinois as an expert who was giving up on varietal crosses.[20] Wallace had not actually changed his views on farmer experimentation with corn; he was speaking instead of the methods used by corn breeders. While farmers should keep working on improving their own corn, breeders were no longer sanguine about the results. In the fall of 1919, for example, Wallace again urged farmers to select corn from their fields at harvest. Although he reiterated his earlier warnings regarding the lack of correlation between scorecard selection standards and corn yield, he nevertheless argued that "to judge seed corn by the conformation of the ear and the kernels seems to be a good deal like guessing the milk production of a cow from her general build. Neither method is sufficient to ensure the highest production, but both are a considerable aid in getting results above the average." Yet only three months before, Wallace had noted how little was really understood about corn breeding, claiming that "from the standpoint of the practical farmer, the situation is not particularly encouraging" and that "the problem of developing higher-yielding strains of corn is so complex that most corn breeders are foredoomed to failure."[21]

---

which is then crossed; "Ear-to-Row Breeding of Corn," *Wallaces' Farmer*, 12 January 1917, 58.

[20]"Illinois Corn Breeding Experiments," *Wallaces' Farmer*, 6 June 1919, 1175, 1187. During the war years, even *Seed World*, the trade journal of the American Seed Trade Association, urged farmers to practice selection and crossing because of a wartime seed scare. See, e.g., Henry G. Bell and Clyde A. Waugh, "Importance of Seed Improvement," *Seed World*, 5 April 1916, 326; D. L. Davis, "Seed Improvement," *Seed World*, 5 January 1917, 28, 30; "Methods of Seed Corn Selection," *Seed World*, 18 July 1919, 50.

[21]"Selecting Seed Corn," *Wallaces' Farmer*, 19 September 1919, 1807; "Illinois Corn Breeding Experiments," *Wallaces' Farmer*, 16 June 1919, 1187.

Thus, for Wallace in 1919 there were two distinct and only slightly overlapping approaches to corn breeding—one for farmers and one for breeders. Farmers who continued to select for higher-yielding corn might actually improve their strains slightly and should be encouraged to keep at it. Breeders, on the other hand, should look elsewhere for improving corn yield.

Why in the space of four years had Wallace shifted from a ringing endorsement to a pessimistic and disappointed abandonment of traditional corn improvement methods? One possibility, offered, I might add, by later breeders, was that selection and varietal crossing were conclusively shown to be not worth the trouble. Some felt that increases of five bushels per acre were simply not economically significant enough to justify the costs of record keeping. Perhaps, as these breeders reported, strains that yielded well one year would not do so the next year or would yield well in one area but not in another. Many felt that varietal crosses were simply too inconsistent.[22]

The more plausible explanation for this shift, however, was that in 1918 Donald Jones had invented double-cross hybrid corn. For commercial corn breeders, this discovery was indeed revolutionary not only for breeding methods but also for the value of seed corn they offered for sale and for the change it predicted in the relationship between farmers and breeders. From Wallace's point of view, farmers should continue selecting corn partly because hybrid corn was still an idea rather than a commodity; Wallace knew that even if it worked, hybrid seed would not reach the marketplace for many years. Moreover, following Hartley's logic that Mendelian breeding methods must rely on the superior corn strains discovered by selection, Wallace felt that farmers could make a real contribution to the hybrid cause by acting, again, as a corps of independent discoverers of those superior lines.

Yet it was clear to Wallace and others that hybrid corn breeding

[22]For example, see D. F. Jones, "Selection in Self-Fertilized Lines as the Basis for Corn Improvement," *Journal of the American Society of Agronomy*, 1920, 12:77–80; Simmonds, *Principles of Crop Improvement*, pp. 140–42.

was "not a practical proposition for the average man." After the creation of the hybrid method, farmers would only occasionally be considered breeders as the gap between breeding and farm experimentation widened. Wallace believed that "the great field for crossing is in rejuvenating in-bred strains. It has been established that crosses of in-bred strains will yield better than either parent two times out of three." Of course, he neglected to mention that the technique of crossing inbreds had been mastered twelve years before and that the big difference in 1919 was that a way had been found to make crosses commercially viable. The method itself had not really changed, but the meaning of it had.[23]

Inventing Hybrid Corn

In 1913 Donald F. Jones asked E. M. East to direct his graduate work in corn genetics. Jones, who was then at the Arizona experiment station working on an alfalfa-breeding project, was especially interested in, though as yet ignorant of, practical corn breeding; he had been galvanized by East and Hayes's 1912 paper on heterozygosis. After teaching horticulture and genetics at the University of Syracuse from 1914 to 1915, Jones was appointed chief geneticist at the Connecticut station, replacing Hayes, who accepted a spot at the University of Minnesota that year.[24]

During his first five years at the Connecticut station Jones developed two ideas that had a profound impact on corn studies. The first, presented in 1917, offered an explanation for the renewed vigor observed in a single-cross hybrid. For several years scientists had been unable to explain why inbreds were deformed, runty, and weak, and second-generation crosses were

[23]"Illinois Corn Breeding Experiments."

[24]Jones, "Biographical Memoir of East," pp. 217–242; Cannon, ed., *Great Men of Modern Agriculture*, p. 225; Crabb, *Hybrid Corn Makers*, pp. 80–92.

considerably less vigorous than the single cross, but first-generation crosses of two inbreds were vigorous and strong. What could account for the special vigor of the single cross; in other words, what did the single cross gain that the second generation lost? Jones rejected the theory advanced by F. Keeble and C. Pellow in 1910 that vigor resulted from a concentration of dominant characters. He argued that complete dominance was in fact unattainable because of a phenomenon he called the dominance of linked factors, in which the physical manifestations of a particular character were dependent on many other, often unobservable factors. Instead, Jones explained hybrid vigor by pointing to the great number of different characters that were combined in the single cross by virtue of what is now called linkage. As Hayes later noted, this work put the phenomenon of hybrid vigor on a Mendelian basis for the first time and created yet another research program for geneticists.[25]

Jones's second idea, which laid the foundation for commercial hybrid corn, was an extension of the linkage hypothesis. Commercial hybrids had remained a fantasy from the time of Shull's single cross because the single cross did not produce enough seed to make it a cost-effective investment for growers. Those breeders with single-cross seed did not know how to transform it into more vigorous seed without losing the hybrid vigor derived from the greater heterozygosity. Jones reasoned that vigor and heterozygosity could both be retained if two single crosses were combined in a double cross, which, while not as uniform as a single cross, would combine all the favorable characteristics of the single crosses, including high yield. Even though the process was initially more cumbersome, requiring the breeder to create and maintain four, rather than two inbreds and to produce two, rather than one cross, the huge increase in seed produced was well

[25]Donald F. Jones, "Dominance of Linked Factors as a Means of Accounting for Heterosis," *Genetics*, 1917, 2:466–479; F. Keeble and C. Pellow, "The Mode of Inheritance of Stature and the Time of Flowering in Peas," *Journal of Heredity*, 1910, 1:47–56; Hayes, *Professor's Story of Hybrid Corn*, pp. 38–39.

worth the effort for commercial breeders. One indication of the economic difference between single and double crosses can be seen in the early price structure of each: single-cross seed was sold in thousand-seed lots, double-cross seed was sold by the bushel.[26]

Jones was not the sort of scientist to sit idly by while others considered the merits of his ideas, and in 1919 he approached Hartley, inquiring "if some arrangement could be worked out whereby as a collaborator with the Department of Agriculture in your department I could meet with these different workers and go over their work with them."[27] And indeed, Hartley had made such arrangements for years with breeders and growers in the states. Most of these collaborators were ordinary farmers who were interested in testing new corn strains and methods, and Hartley paid them either a short-term wage for their trouble or, on occasion, a dollar a year for their general attentiveness to the work.[28] As a land-grant employee himself, Jones no doubt realized that this was a useful connection to cultivate as he took his method on the road. In less than a month Jones was appointed a collaborator "to bring about further cooperation of those engaged in corn improvement work and make a study of corn breeding methods with experiment station workers and farmers," for which he too was paid a dollar a year.[29]

[26]Donald F. Jones, "The Effects of Inbreeding and Crossbreeding upon Development," *Connecticut AES Bulletin 207*, 1918, 5–100; Jones, "Selection in Self-Fertilized Lines as a Basis for Corn Improvement," *Journal of the American Society of Agronomy*, 1920, 12:77–100.

[27]Donald F. Jones to C. P. Hartley, 26 September 1919, 54/31/82, NA.

[28]For example, in 1912 Hartley suggested to Galloway that the bureau hire O. H. Moyer of Augusta, Georgia, "who is a practical farmer and has had some experience in work of this nature. He is conveniently located for our purposes and has expressed a willingness to render the necessary assistance." Mr. Moyer was hired at three dollars per day as a special agent in corn investigations. Hartley to Galloway, 23 January 1912, 54/2/319, NA. James S. Giffen of Ballston, Virginia, was hired as an assistant at a hundred dollars per month for five months, and N. H. Sanford, foreman of the Mount Hope Farm in Williamstown, Massachusetts, was hired at a dollar per year in 1918; Hartley to Galloway, 30 January 1912, Hartley to Taylor, 23 September 1918, ibid.

[29]Hartley to Jones, 30 September 1919, and "Appointment for Donald F. Jones," 54/31/82, NA.

Jones retained no uncertainty about how corn improvement work should proceed. "I have the greatest confidence," he wrote to Hartley, "that selection in self-fertilized lines is the real basis for scientific corn improvement. What little work we have done here along purely practical lines has furnished results so definite and gratifying that it is difficult to restrain one's enthusiasm." Not surprisingly, Hartley had no such difficulty, but he was willing to concede that the BPI's selection work was turning into a disappointment: "While we have in a number of ways been instrumental in increasing acre yields as much as fifteen bushels to the acre and in some cases doubling them, the work has been so scattered that the results have not been outstanding except in the case of a few individual farms." Hartley also suggested that Jones's own state of Connecticut might provide a real demonstration opportunity; since it was not a major corn-growing state, yield increases would be more dramatic and thus it would be easier to interest farmers in corn improvement. Jones was not so sure: "The New England farmer is exceedingly conservative and takes considerable pleasure in selecting his own seed corn. He is therefore not easily induced to take up radically new methods." Jones hoped to find more malleable farmers in the Midwest.[30]

The BPI's collaboration with Jones was not exactly a ringing endorsement of Hartley's corn improvement program. Indeed, Hartley's conciliatory manner with Jones can perhaps be understood in light of a proposal to reorganize corn work in the bureau. Three months before Jones wrote to Hartley, Hartley had received notice from W. A. Taylor that Taylor wanted to merge the Corn Investigations Department with the Cereal Investigations Department under Carlton R. Ball. The idea, according to Taylor, was to coordinate corn work under the auspices of disease investigations. Just two months after Jones was appointed, Ball explained the situation to him: "A rather complete reorganization of the investigation conducted in corn investigations is now under consideration, and as soon as we have progressed far enough I shall wish to write you more fully with reference to our plans and to

[30]Jones to Hartley, 26 September 1919, Hartley to Jones, 30 September 1919, Jones to Hartley, 8 October 1919, ibid.

cooperation in which I shall hope to interest you. . . . I hope to devote our funds and men largely to an investigation of the fundamentals in corn improvement, including genetics, physiology, and pathology, rather than to local problems."[31]

But Jones was not the only one in the USDA eager to abandon traditional breeding methods for more Mendelian ones. In a long letter to Henry Wallace, H. D. Humphrey, in the Office of Cereal Investigations, described the new rationale for inbreeding and hybridizing:

> We feel that there can be no doubt but that by means of inbreeding, we can cause defects to show more plainly and thereby make it easier to eliminate the undesirable. Our experience has lead [*sic*] us to the belief that without cross-breeding, all methods of corn breeding tend to a state of fixedness. As this state is approached, advantages of selection decrease. It then seems necessary to resort to a broad line of breeding in order to create variation and make selection more effective. . . . Our greatest program in corn improvement work must come from a combination of selection and cross breeding. These are the two principal methods of improvement and their tendencies are in opposite directions. Selection tends toward uniformity and cross-breeding [*sic*] tends toward variation. It seems to be the happy combination of these two extreme tendencies that will enable us to improve acre yields of corn.

Indeed, by 1920 the USDA already had a few inbred lines of its own. Wallace had hoped to obtain some but Humphrey reported that they were too late-maturing for Iowa, having been produced in Florida and South Carolina.[32]

The most important USDA breeder in the long run, however, was Frederick D. Richey, who had been appointed assistant in corn investigations in 1911. Richey was well connected to the growing circle of Mendelian corn breeders. He corresponded

[31]Taylor to Hartley, 15 July 1919, 54/2/319, NA; Ball to Jones, 8 December 1919, and Jones to Ball, 10 December 1919, 54/31/82, NA.

[32]H. D. Humphrey to Wallace, 16 March 1920, 54/31/166—Wallace, NA.

with Jones and Wallace, exchanged inbred lines with them, and discussed issues of broad concern to those endorsing hybrids.[33] Indeed, together the three formed a potent lobby for the new method, for each commanded considerable authority within his own context. Jones could speak, to some degree, for both the geneticists and the station workers; Wallace had the support of farmers and private seed houses; and Richey brought the sheer weight and prestige of the USDA to bear on corn questions. Certainly other alliances could, and would, strengthen the arguments of hybrid corn proponents, but it is difficult to name any that so thoroughly reflected the various corn interests in America. And perhaps more to the point, the advocates of traditional breeding methods had none of the prestige or authority of Wallace, Jones, or Richey.

By the early 1920s, the BPI seemed disinclined to accommodate traditional corn improvement programs, preferring to throw its staff and resources firmly behind hybrid development and to abandon what it called Hartley's "local problem" approach. Both C. R. Ball and Richey were impatient with Hartley and missed few opportunities to undermine his efforts. Writing to Taylor in 1921, Ball vehemently criticized Hartley's Bulletin 537, "How to Grow an Acre of Corn," which, Ball claimed, should have been written for practicing farmers but, because of its elementary tone, was obviously aimed at boys. "We much [*sic*] assume that the farmer knows most of the kindergarten facts about farming. To continue to preach popular sermons, telling him just what to do along these lines is likely to discredit the Department and weaken its influence." By inference, Ball was trying to depict Hartley as out of touch with more pressing corn improvement problems. Ball could also be prickly, as when the departmental reorganization was under way. When Hartley would not exchange offices with Richey, Ball personally moved Hartley's belongings out of the office and then complained to Taylor that inside a locked desk

[33]Hartley to Taylor, 8 November 1911, 54/2/319, NA; Richey to Jones, 21 February 1920, 29 March 1922, 54/31/82—Jones, NA.

drawer Hartley was keeping "two fountain pens, two slide rules," and other supplies Ball thought extravagant.[34]

Richey agreed that Hartley was a drag on the system. Wallace asked Richey about the problem directly, and Richey answered that he wanted to fire Hartley "in view of his unwillingness to cooperate and his constant obstructionist attitude of 'passive resistance.' " Richey did not actually fire him, but over the next two years Hartley's position deteriorated even further. In a plaintive letter to his brother, Hartley offered his side of the story: "Things are rapidly getting worse for me in the department and it is weakening my heart action. K. F. Kellerman has had my title of Physiologist in Corn Investigations taken from me unfairly. . . . F. D. Richey was chosen by the Bureau Chief to fix my efficiency rating. He fixed it at 47, 65 necessary for retention in service, and the committee passed it on as fixed by Richey. . . . Richey planned the work so I could not accomplish worthwhile results, and at corn ripening time he personally and maliciously destroyed the records on the hand-pollinated ears, removed some from their rows, preventing me from accomplishing the thing he had planned for me. . . . my reputation is being spoiled and I am in danger of a discharge because of things over which I have no control."[35]

It is nearly impossible now to sort out the accusations and difficulties between Hartley and Richey; in any case, the dispute was soon moot. Hartley's brother, J. W., wrote to Secretary of Agriculture W. M. Jardine to plead his brother's case. Jardine and J. W. Hartley had once belonged to the same farmers' union, and both knew the Kellerman family well. But it was to no avail. J. W. received a prompt reply in which Taylor regretted the "misunderstanding" but felt that no other options were available. "His difficulties have been due, in the main, to his unwillingness or inability to work with others. Each year for the past several years

[34]Ball to Taylor, 21 March 1921, 20 October 1922, 54/2/50, NA.

[35]Richey to Wallace, 4 January 1923, 54/31/166—Wallace, NA; C. P. Hartley to J. W. Hartley, 4 April 1925, 54/2/319, NA.

the value of his services in the Department have [*sic*] decreased, and finally there appears to be no other course open than that now being taken."[36] Hartley was not dismissed, but he was shifted out of Richey's way.

Hartley's decline at the BPI and his ultimate demise could be viewed as merely a generational and personality conflict between him and his colleagues. Hartley may indeed have been a difficult person to work with, especially when, at the age of fifty, he was taking orders from the thirty-six-year-old Richey. If we accept the explanations of others at the BPI, Hartley was simply obstructing the natural development of corn work, and his determined resistance to change was having negative effects on departmental progress. From a more distant perspective, however, Hartley's case seems more a reflection of trends in corn research. If the department was putting its resources into hybrid development, it was doing so both by hiring new breeders committed to that approach and by eliminating breeders who felt that existing corn improvement methods were sufficient. In this light Hartley's departure represented not only the creation of a real consensus on corn breeding in the BPI but a growing belief that the old methods of crop improvement were incompatible with the new. From 1925 on, the BPI dismantled traditional projects and adopted the hybrid model exclusively.

An Emerging Consensus

Even before Hartley's departure, of course, Richey, Wallace, and Jones were absorbed by hybrids. For them, the question of which breeding method to pursue was settled, but many other questions remained. Some revolved around the process of producing hybrids: how could a high-yielding line be identified in

[36] J. W. Hartley to W. M. Jardine, n.d. [ca. 4 to 30 April 1925], Taylor to J. W. Hartley, 30 April 1925, 54/2/319, NA.

the inbred stage; how many inbreds should be grown at once; on what basis could selection of inbreds occur; why did inbred crosses or double crosses perform well in one geographical area but not another? Such problems were hotly and widely debated for years within the hybrid community generally, and their resolution ultimately had less to do with unassailable proof than with consensus.

For his part, Wallace realized after a few years that creating a commercial hybrid would not be easy. Whereas some early hybrids yielded well, many were far poorer than their original parents. In a 1923 yield contest, the best local varieties in Iowa yielded around ninety bushels an acre, the best single crosses yielded about a hundred bushels an acre, and the worst single-crosses yielded a mere twenty to thirty bushels an acre. Faced with such mixed results, Wallace confessed to Jones, "The more I see of this kind of work, the more I am convinced that there are astounding possibilities, while at the same time it may be necessary to go rather slowly before arriving at any very definite conclusions concerning the merits of any particular combinations, because of the fact that combinations of this sort seem to be rather highly specialized." This "fussiness" of hybrids would plague the breeders for many years.

Scientific and technical disagreements about how best to create hybrids could also disguise disagreements about the changing social relations between public and private breeders and farmers. When the BPI dismissed Hartley, ostensibly for his scientific naïveté and resistance to Mendelism, it was also dismissing traditional farmer participation in breeding programs in favor of control by scientists of all aspects of crop improvement. Few farmers could manage to inbreed and cross corn. Unlike corn improvement by continued selection, hybridizing could only be done by those with the time, money, and land to devote to it. Thus, the BPI's choice to pursue hybrids exclusively was also a choice, whether witting or not, to eliminate farmer participation.

Between 1921 and 1925, the fuzzy distinction between the scientific and the social methods of producing hybrids was the source of disagreement between Richey and Wallace. At issue was the role of the federal government in the annual state corn yield contests. Wallace had long been a supporter of the contests, feeling that the publicity and prizes encouraged farmers to take an active role in corn improvement. Wallace also felt that the contests brought to light unusually good strains of corn that might not have been discovered by commercial seed producers. He was particularly interested in finding a high-yielding corn for the north-central corn belt, which, because of its short growing season, had a scarcity of high-yielding lines to choose from.

In early 1921 Wallace suggested to Richey that the BPI sponsor a "practical corn yield contest" in the northern corn belt. The contest, which was to be modeled on one designed by farm adviser Martin Mosher, would emphasize yield rather than appearance by comparing hundreds of different lines grown side-by-side in many different locales. Richey was cool to the idea, arguing that since varieties were such "transitory affairs," there would be little of permanent value to gain. He felt as well that since the future lay with hybrids, there was no reason to waste time and money improving varieties. Wallace persisted, "Of course I am willing to admit that the results of the contest after a few years might prove the whole thing to be absurd. But how can you tell until you try it?" Wallace agreed that "such scientific work as inbreeding, cross-breeding, and different methods of selection" should continue but maintained that breeders also needed to locate promising new seed for such work. "Of course what we are after in these yield contests is to find the men who are growing the high-yielding strains," for how else could northern hybrids be developed but with proven local varieties?[37]

Richey's opposition to the yield contest reflected not only his own strong commitment to hybrids but also his reluctance to get involved in anything that smacked of "local problems." As he waged the battle within the BPI to rid the bureau of old-fash-

[37]Wallace to Richey, 4 March 1921, Richey to Wallace, 11 March 1921, Wallace to Richey, 14 March 1921, 54/31/166—Wallace, NA.

ioned, farmer-oriented crop improvement styles, he was unlikely to support such approaches in the states. Richey could not believe that yield contests would benefit the hybrid cause, in part because he thought that good inbred lines did not necessarily come from good varieties—poor varieties were just as likely to generate good inbreds. Moreover, he felt that the contests attracted commercial seed producers, whose lines tended to be old standards, rather than farmers, who allegedly had better lines. Such contests had rarely turned up unusually good strains, and even if a private high-yielding line were to win, the farmer would not be able to supply the resulting demand. Richey's line of reasoning betrayed his bias. These were legitimate criticisms of the old corn shows, where scorecard corn was the winner, but they were not appropriate to Mosher's yield contest, which was designed as an entirely different phenomenon. Richey's dedication to hybrids and opposition to varietal improvement, however, blinded him to such distinctions.[38]

Finding Richey disinclined to support the varietal yield contest, Wallace began discussing the idea with Richey's superior, C. R. Ball. Wallace wanted the BPI to publish the results of Mosher's 1920 yield contest, which he and Ball agreed had been the best varietal test ever completed, and also to help support the contests with federal money.[39] In a later report to Taylor, Ball claimed that he had surveyed state interest in federal support and received favorable replies from only Illinois and Ohio. Ironically, he reported that although Wallace had requested that Hartley be put in charge of the contests, Hartley was not interested. Ball included a proposal for a yield contest from Mosher himself; by this time the USDA had agreed to publish Mosher's report the following year.[40]

[38]Richey to Wallace, 17 March 1921, ibid.

[39]Wallace to Ball, 5 January 1922, Ball to Wallace, 10 January 1922, 54/31/166—Wallace, NA. As part of his argument that farmers had the better varieties, Wallace noted of the 1921 Iowa Yield Contest, "It is rather striking this year as last year that none of the Commercial Seed Companies were successful in placing." Wallace to Ball, 7 February 1922, ibid.

[40]Hartley seems to have felt that if he were put on this project, he would have to give up supervision of corn investigations in general, which probably was Ball's

Mosher's yield contest presented the BPI with a curious set of choices. As outlined and conducted by Mosher, the contest was meant to indicate which variety of corn was most profitable to grow in each section of the state and "to teach farm people to recognize the most profitable types of corn to select." Mosher was adamant that the contest must be "localized," insisting that "if the purpose as stated is to be accomplished, the work must be on so large a scale and so localized that several men in each locality will be induced to enter their corn in the tests." If farmers could be persuaded to select their corn as recommended over a period of three years, Mosher felt, farmers and seed companies alike would see not only which varieties were superior in any region but how much better corn could be if selection were practiced routinely. It was critical, Mosher believed, to discover not what corn was sold by seed houses, which he felt was unimpressive, but what corn had been privately developed by farmers on their own farms. But the two fundamental features of the contest—farmer participation and enlightened selection—were the very aspects of crop improvement that the BPI was endeavoring to abandon in its own work. Mosher's program was but a more systematic version of Hartley's work in Ohio and other states. From Richey's point of view, the whole effort was out of date.[41]

Under pressure from Wallace, however, the BPI gave its grudging support to Mosher's Iowa contest in June 1922;[42] the authorization memo could hardly have been cooler: "Comprehensive varietal comparisons on a nation-wide basis do not seem justified because of the purely local application of the results," but the department would help out "by insuring the inclusion of certain standard or key strains in such local experiments." In private

intent. Ball to Taylor, 24 February 1922, including Mosher's "Plan for a State Wide Series of Field Corn Tests" (manuscript, 54/2/50, NA).

[41]Mosher, ibid. Mosher's work in Iowa and Illinois during this period was finally published in 1962 as *Early Iowa Corn Yield Tests and Related Later Programs*.

[42]It is tempting to surmise that Wallace's grandfather, Henry C. Wallace, who was then secretary of agriculture, influenced the BPI, although this is purely conjecture.

Richey was more critical, asking Wallace, "Are we warranted in going ahead with expensive comparisons of varieties to determine the better ones for different conditions when we already know the four or five that will give us best results?" For Richey the die had already been cast: if hybrids were the future, why encourage farmers to improve their local varieties? Three years after the test began, Wallace had no regrets: "In my opinion the discovery of several high-yielding strains of yellow corn in northern Iowa has paid for all the cost of the Iowa Corn Yield Test more than a hundred times over."[43]

But by 1925 there were new questions to answer. Most obviously, there was a collection of single-cross hybrids that could also be tested in the yield contest. Again, opinion varied on the wisdom of including them. Richey, who was beginning to see the advertising potential of the contest, felt that entering hybrids would hurt the long-range goal of switching farmers from varieties to hybrids. If the hybrids took all the prizes, he worried, farmers who entered varieties would feel betrayed and the contest would be "put on the rocks just about the time we want to use it to demonstrate the superiority of certain specific crosses." Ball disagreed, suggesting that crosses be made ineligible for prizes. He too considered the ultimate triumph of hybrids certain, but "the scientific idea of the inbreds must be sold gradually, in order that the commercial ideas may be put over effectively when inbreds subject to commercial use are in hand."[44] Wallace agreed with Ball and explained to Richey why the yield contest would support hybrids in the long run:

> The practical corn program, as I see it, is first to build up by means of such corn yield contests . . . the idea that certain strains are good reliable yielders. . . . After this has been thoroughly

[43][Richey?], "Tentative Program for Corn Investigations," n.d. [probably 20 June 1922] (Manuscript, 54/2/50, NA); Richey to Wallace, 10 February 1925, Wallace to Richey, 13 February 1925, 54/31/166—Wallace, NA.

[44]Richey to Wallace, 10 February 1925, Ball to Wallace, 12 February 1925, 54/31/166—Wallace, NA.

> demonstrated by five or ten years of experimenting, we can go ahead and demonstrate conclusively that certain combinations of inbred strains are on the average five or ten bushels per acre better than the best which has heretofore been known. In this way you have the background for putting over in the most definite fashion possible the true value of the new method. At the same time you have brought to the surface farmer strains which should serve as a valuable source of material in inbreeding work.[45]

Interestingly, the Iowa Corn Growers' Association didn't seem to care whether inbred crosses were included in the contest or not, feeling the hybrids that yielded well in the 1925 season were aberrations and would not repeat their performance.[46]

It is perhaps impossible to demonstrate the ability of BPI scientists to reshape research practices in the states. Nevertheless, it is clear that research at experiment stations underwent a pronounced shift after about 1925. Until that time, stations all over the country reported varietal testing and selection experiments as well as Mendelian studies of inheritance. After 1925, however, nonhybrid studies almost entirely dropped off the roster of experiments at most stations. Those stations performing field trials and other traditional tests now did so with inbreds or single crosses rather than open-pollinated strains. The new era had begun.[47]

[45]Wallace to Richey, 13 February 1925, and see Richey to Wallace, 20 February 1925, ibid. Wallace finally tried to end the dispute: "I guess it is difficult for us to disagree so very much even when we try as hard as we can. About the only difference I can see in our viewpoints now with respect to corn yield contests, is that I think that highly productive farmer strains brought to the surface by the yield contests may have real genetic value for our future inbreeding work. You, on the other hand, apparently think that in all probability any ordinary strain will be just about as good for inbreeding work as a strain which has proved its superior yielding power in a corn yield contest. In the main, just judging from the theory of probability, I know that you are right. And yet there is always a remote chance that some of these "dark horse" farmer strains will contribute something genuinely worthwhile." Wallace to Richey, 24 February 1925, ibid.

[46]Wallace to Richey, 13 February 1925, ibid.

[47]The change in corn research during this period was chronicled in the *Experiment Station Record*, which abstracted station bulletins and reports. As I will

Why Hybridize Corn?

The BPI's minimal support of the Iowa Corn Yield Contest did little to stem the tide of enthusiasm for hybrid corn, either within the bureau or outside it. Richey continued to encourage hybrid studies and to dismantle varietal testing programs. Writing to George Sprague, then in North Platte, Nebraska, Richey announced that "the experiments on varietal crosses may well be discontinued" because "crosses between self-fertilized lines seems [*sic*] to be a more certain method of obtaining increased yields." A 1924 annual report from the Office of Experiment Stations did not even discuss selection as a method of crop improvement, simply noting that selection work continued in the Illinois high-low oil and protein study, which by then was considered more a curiosity than a promising methodology.[48]

The actual "success" of hybrid corn a mere ten years later does little to explain why the BPI abandoned traditional methods in the early 1920s. At that time hybrids were not an obvious or direct path to corn improvement. Breeders left some questions unanswered. First, why did they think it necessary to increase corn yield so dramatically? It is true that disease, especially in the midwestern states, affected yield somewhat adversely, but this problem was being attacked by the traditional method of educating farmers in selection techniques—and, as I show in a later chapter, with considerable success. In any case, despite losses from disease, there was no chronic shortage of seed corn during this period. Although the occasional unseasonal freeze or dry summer could reduce seed supplies in some areas in some years, the problem was neither persistent nor widespread enough to

---

discuss later, this change in research did not necessarily signify a change in field practice, at least until the mid-1930s.

[48]Richey to George F. Sprague, 17 July 1925, 54/31/134—Sprague, NA; Henry M. Steece, "Breeding Work with Field Crops at the Experiment Stations," *Office of Experiment Stations Annual Report*, 1924, 51. See also Richey, "Effect of Selection on the Yield of a Cross between Varieties of Corn," *USDA Bulletin* 1209, 1924.

warrant an all-out national effort to increase the yield of corn.[49] It did not require a Ph.D in economics to realize that higher yields were likely to lower prices, but breeders, for the most part, were silent on this issue. Whether the discussion took place in the popular press, USDA publications, or private correspondence, the assumption that more corn was a good thing was ever present and uncontested.

Although proponents of the selection method were likewise interested in increasing corn yield, their assumptions and aims stood in marked contrast to those of the hybridizers. First, although increases in yield from selection could be quite large, in general they tended to be moderate, that is, large enough to justify the effort but not so large as to interfere with the general economy of corn production. There seemed to be a sort of natural ceiling on improvement that could be accommodated by existing markets. Second, the selection method depended on the active participation of farmers and encouraged their habit of saving seed from one year to the next. If a farmer was trying to build up a supply of good corn over the period of a few years, which is how selection worked, then there was little reason to buy fresh, unpredictable seed each year. For seed dealers, selection was decidedly a step in the wrong direction. Learning how to systematically select corn kept farmers in control of the crop in a way that buying seed did not. As Jones had pointed out, many farmers were quite proud of their ability to increase their own yields by selection. They knew exactly what their corn was and were not at the mercy of seed dealers, who made claims that might or might not be true. Farmers who selected their own corn year after year could take all the credit—or all the blame—for the results.

It would seem, in fact, that increasing the yield of the corn crop

[49]Weather-related seed shortages hit the Midwest in 1918 and again in 1924. See "The Seed Corn Situation," *Seed World*, 5 January 1918, 18; "Posters to Help Seed Corn Situation," *Seed World*, 5 February 1918, 89; J. C. Hackleman, "Selection and Storage of Seed Corn," *Seed World*, 19 September 1924, 13–14; "Summary of Seed Corn Situation," *Seed World*, 7 March 1924, 13–15, 28.

was not the compelling reason why breeders from the BPI and elsewhere pursued hybrids so single-mindedly. Other aims carried far more weight. First, American breeders in general and the first generation of plant geneticists in particular were obssessed with Mendelism. Mendel's laws not only provided them with a way to direct plant improvement but defined for them a methodological and professional niche that had not existed before. For breeders like Richey, use of the Mendelian method distinguished the new scientific breeder from the old rule-of-thumb breeder. It was an exciting and powerful research program that promised the twin results of legitimating breeding as a scientific enterprise and revolutionizing the mundane practice of crop improvement. Following the logic of "if it can be done it should be done" characteristic of so many innovative techniques and ideas, such breeders found Mendelism irresistible, whatever the consequences.

Second, commercial seed producers had long bemoaned their inability to control corn varieties. Many felt that if they went to the trouble of developing and maintaining a particular line, they should be paid accordingly, and as a group seed producers opposed the idea of farmers saving their own seed. After the seed shortage of World War I, for example, they protested when the USDA continued telling farmers to save seed for planting. Claiming that farmers were not really capable of producing good seed themselves, commercial producers tried to think of ways to stop such practices. As one wrote, "There is no reason why the business of legitimate seedsmen should be interfered with and farmers and gardeners urged to produce their own." In essays titled "Beating the Farmer at His Own Game" and "Discourage Home Seed Saving," seed producers expressed their growing frustration.[50]

Occasionally a breeder would try to get a patent on a corn line,

[50]"Home Seed Saving Again a Factor," *Seed World*, 15 August 1919, 22–23; "Beating the Farmer at His Own Game," *Seed World*, 18 April 1919, 512; "Discourage Home Seed Saving," *Seed World*, 4 April 1919, 422.

to no avail. Donald Jones was especially keen on the subject of corn patents. As early as 1922, he began considering how to protect "methods of producing hybrid seed." He first broached the topic with Ball: "The method of separation by the use of endosperm and aleurone colors may also be subject to protection by patent and if so, it would be desirable to have these patents kept out of private hands." Richey disagreed, arguing that any "patentable" method "should be protected for the benefit of all people," and in any case, according to Richey, it would be difficult to obtain a patent on detasseling or hand-pollinating because these techniques had been used commercially for years.[51] Though Jones was ultimately disappointed in his efforts to patent hybrids in the 1920s, he renewed his claim in 1948 by filing a patent with Paul Manglesdorf. This effort also failed, but Jones continued filing patents, and it was not until 1970 that all such claims were abandoned.[52]

I do not mean to suggest that the BPI chose to "side with" the commercial breeders and seed producers against farmers. There is no clear evidence that the BPI viewed the situation in such terms. Moreover, farmers and producers were not discrete and separate entities; many seed producers were farmers, and of course farmers could be "old-fashioned" or "progressive." Farm-

[51]Jones to Ball, 8 November 1922, Richey to Ball, 11 November 1922, Ball to Jones, 11 November 1922, 54/31/82—Jones, NA; Ball to Jones, 4 May 1923, 54/31/166—Wallace, NA. Ball was a little worried about patenting the hybrid method, recalling the case of a Texas farmer who received a patent on the "method and time of plowing" and then tried to collect royalties from other Texas farmers; see Ball to Jones, 1 June 1923. Jones responded with his own example of a Massachusetts experiment station that had developed a new tree surgery procedure. When a commercial firm patented the procedure, the station found that it was no longer allowed to use the method it had actually invented. Ball responded that he "can not conceive of" such a situation; see Jones to Ball, 9 May 1923, and Ball to Jones, 12 May 1923, 54/31/82—Jones, NA.

[52]The story of Jones's later patenting efforts has not yet been told, although Stanley Becker drafted an unpublished account of the proceedings. I am grateful to Diane Paul for bringing this situation and Becker's draft to my attention. All pertinent materials are located at the Research Corporation in Tucson, Arizona.

ers, in any case, do not seem to have perceived themselves as constituting an interest that would be threatened by hybrids. Rather, the point here is that the BPI chose to support the hybrid method and to abandon the selection method, and this choice had the effect, desired or not, of eliminating the role of most farmers in the corn improvement process. Just how well this decision played in the states remains to be seen.

3

Building the Machine:

Development of the University of Illinois College of Agriculture

The disputes between scientists over which forms of corn breeding were appropriate or practical underlined an essential truth about agricultural science: in differing agricultural contexts, scientific ideas may flourish or expire depending on the social arrangements that obtain among scientists, farmers, and administrators. Shull, who had little firsthand experience with farming or farmers, felt that what could be done, should be done. East and Wallace, much more familiar with the concerns of farmers and seed producers, knew that it was not so simple. The tenuous relationship between expert knowledge and farming needs, between scientists and farmers, was neither so secure nor so malleable as Shull seemed to assume.

In this chapter I discuss the context of agricultural science in Illinois, focusing particularly on the agricultural experiment station, the extension service, and farmers' organizations. From the beginning the effectiveness of the station and the extension service explicitly relied on support among Illinois farm groups. Not surprisingly, these programs were legitimated by the argument that agricultural science was largely an applied discipline, meaningful only within the confines of farming concerns. Both the station and extension were designed in cooperation with farm interests and welcomed the active participation of farmers and farm businesses. Yet, as will become apparent, the "farm interests" in Illinois were in fact a widely heterogeneous collection of small and large farmers, agricultural businesspeople, and lobby groups, who rarely spoke with one voice. Further, when the college attempted to act in the "farmers' best interest," it was not always clear which of these groups would in fact be served.

These interrelationships between the college and its clientele provided opportunities for research and education, as well as constraints on them. On the plus side, the participation of farm groups in defining research problems kept college and station scientists in touch with regional farm problems and provided the college with a group of farmers willing to test new ideas and techniques on their own farms. On the other hand, the sometimes conflicting demands of small farmers and agribusiness, for example, repeatedly forced the college into compromises that satisfied no one. Gradually, these frustrations and limited resources led the college to distance itself from its clientele. Faced with an insistent and powerful competitor to its basis of authority—agribusiness—and the growing complexity of both scientific and political issues, the college retreated.

Before I can examine how the different corn-breeding alternatives were adopted or abandoned in different contexts and, in particular, how the University of Illinois dealt with the changes, it is necessary to consider how the university's options were shaped by its historical relationship with the agricultural interests in the state. Just as East and Wallace knew what tacit agreements they had made with their farm clientele, so scientists and administrators at Illinois understood their own prerogatives in terms of which ideas their supporters would accept or reject. Perhaps few university scientists in the 1930s could remember how the curious alliance between farm and college had been formed by Eugene Davenport, dean of the Agricultural College from 1895 to 1922. But few could escape his legacy.

Davenport and the Agricultural Experiment Station

Eugene Davenport (1856–1941) was born and raised in rural Michigan and attended Michigan Agricultural College, graduating in 1878. He worked on his father's farm off and on for the next ten years, taking time off to obtain a master of science degree in 1884. From 1888 to 1891 he taught agriculture at his alma mater, first as assistant to botanist W. J. Beal, who did pioneering work in corn breeding, and later as professor of agriculture and superintendent of the university farm. In 1891 he left the university with few regrets after agreeing to become president of a proposed agricultural college in Brazil. After only six months, however, both the college and Davenport's position there were aborted when the political climate proved inhospitable. Davenport returned to Michigan by way of England, and remained there until early 1895 when he accepted the deanship at Illinois.[1]

Illinois Industrial University (renamed University of Illinois in 1885) was founded in 1867 "to promote the liberal and practical education of the industrial classes." In its early years, the university was nearly ruined by curriculum disputes, meager income, poor enrollment, and popular disdain for its goals. The agricultural program, moreover, suffered for lack of faculty and from the unreasonable expectations of the administration. Before Davenport arrived in 1895, agricultural experimentation had been conducted only intermittently. In 1870 Thomas Burrill and Willard Flagg investigated field practices, soil fertility, varietal differences in grains and vegetables, and fruit diseases. In the late 1870s and 1880s Thomas Hunt continued the field experiments and began livestock-feeding studies; George Morrow studied corn production practice on his now famous Morrow plots; and Henry Weber and M. A. Scovell conducted soil analyses and worked with sorghum.[2]

This experimental work went on in facilities whose wretched state shocked Davenport. There was no agriculture building, the barn was dilapidated and empty, and equipment was virtually nonexistent. More alarming still was the lack of course offerings and students. In 1895 there was one winter course and only nine

[1]Moores, *Fields of Rich Toil*, pp. 108–109. Moores's history is an excellent study of the development of agriculture at Illinois.

[2]Moores, pp. 3–99 passim.

of the eight hundred university students were enrolled in agriculture. In part the enrollment problem reflected the general underdevelopment of rural education. Since rural secondary schools in Illinois were few and far between, most farm children were unlikely to take the university entrance examination and were even less likely to pass it. Furthermore, those few students who were admitted had to wait until their third or fourth year before beginning agricultural studies; by that time most had become interested in something else or had become discouraged and left the university.[3] Moreover, many farmers were skeptical about the advantages of attending the agricultural college, and some apparently felt that their sons could do better by following a career other than farming and so discouraged them from attending the college.[4] Even the president of the university, Andrew Sloan Draper, was indifferent to the College of Agriculture and repeatedly fought Davenport's efforts to improve the college and the station.[5]

Davenport's response to these circumstances reflected his belief that "leadership must be with the farmers not the university," and during the next ten years he concentrated on building up the agricultural college by enlisting the support of farmers and farm organizations. In addition to making public appearances before farm groups, Davenport worked behind the scenes with powerful agricultural leaders in the state and helped organize farmers into associations that could represent them before the legislature.[6]

Davenport concentrated his earliest fund-raising efforts on the county-based Farmers' Institute in Illinois, which, like those in other states, was loosely made up of successful farmers and rural

[3]Davenport, "Rejuvenation of the College of Agriculture of the University of Illinois, 1895–1922" (manuscript, November 1933, University of Illinois Archives [UIA]).

[4]W. J. Kerr, "Some Land-Grant College Problems," *Association of American Agricultural Colleges and Experiment Stations (AAACES)*, 1911, 41.

[5]Moores, pp. 101–149. In 1898 seventeen Illinois farm boys were studying agriculture in out-of-state colleges; see Davenport to Colonel C. F. Mills, 21 October 1898, Davenport Letterbooks, UIA.

[6]Davenport, "Rejuvenation," p. 52; Moores, pp. 101–149.

educators, who wanted to take the lessons of progressive agriculture to local farmers. Unlike the Farmers' Alliance and the Grange, it was pedagogic rather than political. In Illinois institutes sponsored itinerant lectures and mildly educational meetings around the state. In 1869 Illinois Industrial University launched a series of such lectures, but farmers' response was uniformly disappointing and the idea was abandoned after four years. Nonetheless, county institutes had kept the flame alive in varying measures, and they appeared to Davenport the logical starting point for his campaign.[7]

In 1895 Davenport tapped into the institutes through Colonel Charles Mills, secretary of the State Board of Agriculture, who had been a prime mover in organizing county institutes in Illinois. By 1894 nearly half the state's counties could claim a Farmers' Institute association. Mills and Davenport felt that a more organized state association of farmers would constitute a powerful lobbying group before the legislature. If properly motivated, these farmers could argue more persuasively than educators that agricultural progress in Illinois depended upon the state's generosity to the agricultural college. With this idea in mind, Mills began agitating for such an organization, and in 1895 the Illinois Farmers' Institute was created by the legislature as a permanent association "to assist and encourage useful education among farmers and for developing agricultural resources of the State."[8] The state organization was composed of three delegates from each county, plus a board of directors consisting of the president of the Illinois Horticultural Society, president of the Illinois Dairymen's Association, director of the State Department of Agriculture, dean of the College of Agriculture, and the state superintendent of public instruction.[9]

Although the institute would not receive state funding for several more years, the machinery was in place for cooperation

[7]Roy V. Scott, *The Reluctant Farmer: The Rise of Agricultural Extension to 1914* (1970), pp. 79–80.

[8]Davenport, "Rejuvenation," pp. 34, 31.

[9]Ibid.; Moores, pp. 111–114.

between the institute and the agricultural college. In time, the institute became a particularly compelling voice for legislative support of the college and especially the experiment station. And it appears that state funding did hinge to a large extent on farmers' active endorsement of the station agenda.[10]

But in the early years farmers' support for scientific agriculture was hard won and focused more on the experiment station than on the college. This distinction was brought home to Davenport in his first successful legislative battle in 1898. In 1895 and again in 1897 the legislature had rejected college requests for an appropriation to construct a dairy building on campus. Farmers, for reasons that are unclear, were opposed to the appropriation. A likely explanation is that dairy interests were neither sufficiently large nor well organized enough to push the request through, and the majority of farm interests were opposed on general principles to spending state funds on what they perceived as specialized agricultural education. In 1898 Davenport requested funds for a building devoted to agriculture rather than dairy and persuaded these activist farmers that the agriculture building was in their best interests because it would allow a more extensive investigation of pressing farm problems. The institute not only supported Davenport's idea but raised the request from $100,000 to $150,000.[11]

In the same year the Farmers' Institute joined Davenport in persuading the legislature to redistribute federal land-grant funds in a way that would increase the station's allowance. Until then the university as a whole had received $56,000 per year from these sources but only $7,000 went to the station. Davenport proposed that the money be evenly split among the agriculture, engineering, and liberal arts colleges. But the legislature, feeling suddenly expansive, divided the money between the agriculture

[10]In 1902 the USDA formally recognized the Farmers' Institute and gave the Office of Experiment Stations two thousand dollars to promote its efforts. See Gladys Baker, *The County Agent* (1939).

[11]Moores, pp. 119–120; Davenport, "Rejuvenation," pp. 35–40.

and engineering colleges and threw in an equal amount for liberal arts.[12]

But even as the legislature was becoming interested in Davenport's plans and persuaded of its obligation to support state agricultural work, both bills were threatened by enemies within. Directors of the institute, apparently angered because the state gave ten thousand dollars more to the rival engineering college than to the agricultural college, announced that once the money was received, the institute would remove the experiment station as well as Dean Davenport from the university. They were not interested in education and wanted to guarantee that their production problems would be addressed without compromise and that the control of such matters would rest with them. A stunned Davenport managed to defuse this plan, but it was clear that while the farmers had almost too enthusiastically adopted his experiment station ideals, their feelings toward the agricultural college were considerably less warm.[13]

Furthermore, even though the legislature approved the redistribution of funds Davenport had suggested, university trustees decided not to implement the decision. Indirect evidence suggests that President Draper originated this scheme. Davenport confronted the trustees at a meeting notable for both the absence of Draper and the presence of Farmers' Institute members (albeit in the hallways) and managed to persuade them to approve the appropriation.[14]

In November 1900 Davenport again turned to the farmers for help in approaching the legislature for station funding. Together Davenport and the farmers drew up a budget proposal for scientific investigations, including $25,000 for livestock, $10,000 for corn, $10,000 for soil, $10,000 for orchards, $10,000 for dairy, and $5,000 for sugar beets.[15] Reasoning that each each particular

[12]Davenport, "Rejuvenation," pp. 42–46.

[13]Ibid., pp. 37–38; Moores, pp. 121–123.

[14]Moores, pp. 132–135.

[15]According to Davenport, "There were positively no sugar beet interests in the state when that section was drawn. But agents were canvassing the territory in a

request would fare better with a sponsor, he helped create three new associations: the Illinois Live Stock Breeders' Association, the Illinois Corn Growers' Association, and the Illinois Beet Growers' Association. The Farmers' Institute agreed to sponsor the soils research portion of the bill, and the Horticultural Society and Dairymen's Association agreed to sponsor their respective sections. Each association appointed an advisory committee that worked directly on Bill 315, as it was called, thus presenting a united front of agricultural interests before the legislature. Not surprisingly, Bill 315 passed in 1901, and although livestock, dairy, and beet appropriations were less than requested, the $54,000 appropriation was a triumph for Davenport, who felt that his mandate was at last taking shape.[16]

The commodity groups Davenport created, moreover, were retained as advisory committees to the College of Agriculture. In 1901 they replaced the advisory board (also called the Board of Direction), which for several years had assisted Draper and Davenport in setting agricultural agenda at the college. Because the committee members were largely drawn from among the most successful farmers in Illinois, who not only had practical farming experience but also some degree of social and political influence, the advisory committees were quite effective in directing Daven-

---

campaign to sell stock for the erection of sugar beet factories before anybody knew whether Illinois was fitted to grow beets or whether Illinois farmers cared to diversify in this direction" ("Rejuvenation," pp. 50–51). The corn breeder Perry Holden left in 1900 to work in such a factory but quit after one year.

[16]In addition to the Corn Growers' Association, the Illinois Seed Corn Breeders' Association was formed during the same period. It appears that the growers were concerned primarily with sponsoring corn shows, and the breeders were more interested in breeding and selling "purebred" seed corn. The distinction between the two is by no means clear, however; some prominent breeders, such as Eugene Funk, E. E. Chester, and H. S. Winter, were also growers, while other breeders, such as L. F. Maxcy and C. A. Rowe, were not. Official documents of the two associations are neither plentiful nor helpful; see "Constitution and By-Laws of the Illinois Seed Corn Breeders' Association," May 1911, which includes a list of members and officers (Alexander Collection); and "Illinois Corn Growers' Association First Annual Report," 1902 (Funk Brothers Seed Company Files).

port's attention to practical and, in particular, economic considerations.[17]

And indeed, in 1901 the agricultural program at the college looked promising. The agriculture building was completed, and the college began offering agriculture scholarships to boost enrollments. Working again through the Farmers' Institute, the college gave agriculture fellowships to one student from each county who was recommended by the institute, thus swelling the agriculture college enrollment from 19 in 1898 to 159 in 1901. In addition, the station was nearly overwhelmed by farmers interested in testing and experimentation, which "captured the imagination of the public as nothing else agricultural had done in the history of the agriculture colleges." In this period, according to Davenport, the university as a whole was most commonly referred to as the station because "the needs for results of research were more easily understood and of more immediate interest than was the problematical development of instruction."[18]

Nevertheless, the situation had improved dramatically. Davenport had begun in 1895 with a skeptical and disgruntled farmer clientele, no facilities or equipment, no student body to speak of, and a poorly funded station and college, which lacked the support of both farmers and legislators. Even the university president was against it. By arguing that the agriculture college belonged to the farmers, that they had a right and an obligation to direct its course, Davenport created a support system that time and again compelled the legislature to appropriate necessary funds for the college and the station. Writing to A. B. Storms, president of Iowa

[17]"Notes on the History of the Advisory Committees" (manuscript, 22 March 1923, 8/1/2, box 17, UIA); Moores, pp. 146–147; *Proceedings of the Board of Trustees*, 1901–2, passim. The outgoing board in 1901 consisted of J. Irving Pearce of the Illinois Board of Agriculture, H. Augustine of the Illinois Horticultural Society, H. B. Gurler of the Illinois Dairymen's Association, S. Noble King of the Illinois Farmers' Institute, in addition to agricultural faculty and university trustees.

[18]Davenport, "Rejuvenation," pp. 41, 16, 53.

State University, who asked Davenport's advice as to the wisdom of aligning himself with agricultural associations, Davenport reported that such alliances were very useful in obtaining money with which to conduct research.[19]

But by 1905 the situation took an unanticipated turn when it became clear that success, too, had its cost. Davenport had skillfully implanted the station concept in the minds of Illinois farmers, and it was the station, rather than the college, that represented to them the virtues of scientific research and agricultural progress. While Illinois legislators and farmers were keen to support research, their generosity created problems for the rest of the college, and in 1904–1905 the college was in distress. The 1901 appropriations and expanded student body placed an impossible burden on faculty and educational facilities. Although Davenport was almost solely responsible for this state of affairs, he later complained that "instead of recognizing this response to research as the first sign of a possible interest in agriculture and a foundation for the college of agriculture, the institution practically sold out to the experiment station, and as to college work, went practically out of business."[20] As before Davenport rallied the organized farmers, who successfully appealed for legislative aid, and the crisis was averted, but the station appropriations continued to surpass those for the college.[21]

By 1910 university officials realized that a more permanent and rational source of college funding was vital. In August, Davenport assembled the six advisory committees to evaluate the needs of the college, and the group quickly agreed that "conditions were most critical and the existence of the Agricultural College as a

[19]Davenport to A. B. Storms, 22 June 1905, 8/1/2, box 1, UIA.

[20]Davenport, "Rejuvenation," p. 16.

[21]Moores, pp. 146, 149. Between 1905 and 1911 the biennial state appropriations for the station increased from $190,000 to $276,000, while those for the college actually fell from $122,000 to $110,000. And whereas the college and station jointly received $415,300 for the 1911–13 biennium, the station got an extra $336,000 for research alone.

school of the first rank was at stake." The primary problem, as they perceived it, was that the college had not expanded to accommodate the growing student body, and research had not kept pace with the needs and interests of the state. Ten days after this meeting, a subcommittee set out to investigate conditions at agricultural colleges in other states with the intention of learning how to expand most efficiently and economically. More to the point, the committee hoped to embarrass the Illinois legislature by demonstrating that other states took their responsibilities toward agriculture more seriously and so to galvanize the legislature into approving sufficient funds. And indeed, the committee collected statistics that seemed to verify their point. Claiming that Cornell University was "strictly comparable with Illinois," the committee showed that for a similar enrollment Cornell had twice as many teachers and spent twice as much money on the college. Of the nine schools visited, Illinois ranked last both for funding devoted specifically to the college and for value of livestock.[22]

The report of the agricultural committee was presented to the Illinois legislature early in 1911 in the form of a bill, requesting $670,750 for buildings and equipment, $574,000 for salaries and maintenance, and $326,000 for station research. To ensure that legislators gave the proposal serious consideration, a subcommittee was appointed to lobby on its behalf. The group included Frank Mann, auditor of the Illinois Farmers' Institute; Ralph

[22]"Report of the Agricultural Committee on the Needs of the College of Agriculture of the University of Illinois" (December 1910, 8/1/2, box 20, UIA), pp. 2, 13. Members of the committee were J. R. Fulkerson (Illinois Live Stock Breeders' Association); G. D. Montelius and S. W. Strong (Illinois Grain Dealers' Association); C. A. Rowe and L. F. Maxcy (Illinois Corn Growers' Association); E. W. Burroughs and H. A. McKeene (Illinois State Farmers' Institute); R. O. Graham and W. G. Lloyd (Illinois Horticultural Society); T. Lamb and J. A. McCreery (Illinois Grain Dealers' Association); A. N. Abbott and W. Huffaker (Illinois Corn Growers' and Stockmen's Convention); C. L. Washburn and J. F. Ammann (Illinois State Florists' Association); L. F. Maxcy and H. J. Sconce (Illinois Seed Corn Breeders' Association); and L. F. Maxcy and A. W. Jamison (Illinois Farmers' Club).

Allen, director of the Farmers' Institute; Harvey Sconce, corn breeder and farmer; C. A. Ewing, lawyer and "farmer in a large way"; and W. N. Rudd, president of Mount Greenwood Cemetery in Chicago and "identified with the ornamental branches of horticulture." When the legislature finally voted on the bill, the appropriation for buildings was slashed to $175,500; nonetheless, the total amount approved for the 1911–1913 biennium exceeded the university's total income for its first twenty-seven years.[23]

It appears that the college and its supporters were not the only ones tiring of the chronic financial plight of the agriculture college and the divisive squabbling over which division was more deserving of funds. At its 1911 meeting, the Illinois legislature passed a bill providing for a "one-mill tax for each dollar of the assessed valuation of the taxable property of this state . . . to be set apart as a fund for the use and maintenance of the University of Illinois." With this, the state appropriation was presented in a lump sum, forcing the experiment station and the agricultural college to negotiate the expenditure between themselves. Thus in one fell swoop the legislature created a permanent base of support for the agricultural college and disengaged itself from disputes within the university. Although this appropriation was no doubt made largely in response to the repeated lobbying of farm groups affiliated with the college, it should be interpreted as well as the successful culmination of Davenport's efforts to use the station as a rhetorical device for obtaining farmer and legislative support. By 1911 the station had attained a permanent role in Illinois agriculture, and it was no longer necessary to rally support anew each fiscal year. Davenport could turn his attention to the crisis resulting from station popularity: the burden of extension.[24]

[23]"Report of the Committee," pp. 2–3; Moores, pp. 146–149.

[24]Moores, pp. 240–241; Edmund J. James, *Sixteen Years at the University of Illinois, 1904–1920* (1920), pp. 27–28.

The Politics of Compromise: Extension

By 1910 it was apparent to agriculturalists that the existing structure and organization of land-grant colleges was painfully outdated. Speaking before his colleagues at the Association of American Agricultural Colleges and Experiment Stations meeting in that year, Davenport stated the problem: "The college was established to teach prospective farmers the science of agriculture, and the stations were speedily found necessary in order to develop that science. We all deplored the long period of apathy on the part of the public; but it was far less dangerous than the present period of universal popularity and unlimited demands upon these new, and as yet partially developed, institutions of higher learning."[25] Davenport's campaign to interest farmers in the station had been all too successful. He had, in effect, transformed the station into a de facto extension service by convincing farmers that research was useful to them. Between 1900 and 1910 farmers had become so keen to take advantage of college and station expertise that Davenport was unable to handle the volume of inquiries; the staff spent so much time responding to farmers that there was little time left for research.[26]

The extension service at Illinois developed largely as a response to these growing demands from farmers for advice and information, but it was also a mechanism for sustaining farm support. In 1901 Davenport appointed Fred H. Rankin to represent the agricultural college in liaison with the Farmers' Institutes. Although his duties were initially limited to corresponding with institute members, by 1903 Rankin was spending a good deal of time away from the college. He was designated college

[25]Davenport, "Shall We Ask for Further Legislation in the Interests of Agriculture? If So, What?" *AAACES*, 1910, 79.

[26]In 1910 the university received $10,700 for extension ($2,700 of it from the state), roughly one-tenth of Iowa's appropriation. Funding varied dramatically from state to state: Alabama State Agricultural and Mechanical, a black school, received only $31 (*AAACES*, 1911, 88).

extension superintendent, and under his leadership, extension became more than a source of information; it assumed the larger role of encouraging farmers and their families to participate in college-sponsored programs and projects. For example, to induce farm boys to attend the agricultural college, Rankin carried on an extensive correspondence, established experimental corn-growing clubs, and arranged excursions to visit the university.[27]

Davenport had somewhat mixed feelings about these outreach efforts. While the college owed a debt of service to the farm community and indeed had rallied farm support by claiming to have broad and useful knowledge, the extension responsibilities of the college were unclear. Davenport had little patience with complacent farmers and aimed his work instead at those who were more "progressive," declaring, "I do not agree with the proposition that the college . . . should . . . take the message to every individual, on the principle of letting no guilty man escape. There will always be lost souls in farming . . . and there will be men not worth saving; for this is public business and not charity."[28]

If Rankin's extension efforts had been the sole strain on the college, the problem may have languished there, but the station was deluged with demands that were, according to Davenport, "varied, many of them erratic, not to say absurd." Farmers wrote

[27]A. C. True, *A History of Agricultural Extension Work in the United States, 1785–1923* (Washington, D.C.: USDA Misc. Publ. 15, 1929), pp. 45–46; Fred Rankin to Davenport, 24 November 1903, Agricultural College Extension, 1903–1921, file 342, UIA. While early extension work may appear quaint, the boys and girls clubs in particular became extremely effective in later years. In 1919 the net profit on boys and girls projects exceeded $127,000 (J. H. Greene to Davenport, Annual Report of Club Work, 6 February 1919, ser. 8/1/5, UIA). The soil extension work was less successful. Meetings tended to attract "mixed" audiences of "town people only indirectly interested in agriculture" (J. E. Readhimer to Davenport, 31 December 1919, file 313). As far as increasing college enrollment, by 1929 officials admitted that extension had failed; see H. W. Mumford, "The Influence to Date of Smith-Lever Extension Work on Rural Life in the United States," *AAACES*, 1929, 256–263.

[28]Davenport, "Shall We Ask," p. 78.

for advice, requested bulletins, made suggestions, and asked for soil and seed analysis, straining staff and financial resources and interfering with research.[29]

Davenport was not alone in feeling that the station and college were being overwhelmed by the expectations of farmers. Station directors across the country were likewise feeling the pinch. They turned to the Association of American Agricultural Colleges and Experiment Stations not only to voice their frustrations but to request more federal aid. By and large these administrators looked to the government with some trepidation. Since 1906, when the USDA had established the Office of Farm Management to coordinate farm demonstrations and county agent work in the northern states, many station directors had grown to resent the federal government's style of cooperation. Davenport in particular chafed under what he considered the USDA's authoritative and ignorant direction of local problems in the states.

In his critique of the Smith-Lever Bill, then pending in Congress, Davenport warned his colleagues of the dangers inherent in acquiescing to the USDA's proposal for cooperation: "If favorable opportunity be afforded by the creation of the necessary machinery, it will be only a question of time until some Secretary will take advantage of the situation to make the state institutions little else than substations of the Department." The question was not whether the federal government would provide funds for state extension programs but whether the USDA would take over state work should such funds materialize. For Davenport and others, the only response to a continued USDA presence in the states would be "to surrender their federal funds and depend entirely for support upon their state constituency, which in the long run pays the bills in either case."[30]

Passage of the Smith-Lever Act in May 1914 in no way simplified the administration of the college; indeed a great deal of

[29]Ibid.
[30]Ibid., p. 7.

confusion attended administration and program development in the early years.[31] At Illinois, Davenport effected a compromise arrangement that more or less satisfied both his agenda and that of the USDA. He was not in favor of boys' and girls' clubs and refused to contribute funding for them but recognized that "if we do not enter into cooperation [the USDA] will withdraw from the state, leaving a rather ugly situation for us to explain." He was also less than enthusiastic about home economics demonstrations but agreed to cooperate if the USDA paid for the whole thing, as he required it to do with club work. Davenport was more enthusiastic about demonstrations in farm management, figuring he could "make out of the whole farm advisers scheme a kind of laboratory in farm economics." He also supported county demonstration work in counties that evinced sufficient local interest and funding, and he agreed to maintain both a state extension leader and departmental advisers to county agents.[32]

The extension legislation in 1914 also created problems for the experiment stations. In 1905 the favorite son of farmers and legislators alike had been the station; by 1919, however, it was the extension service. E. W. Allen of the USDA complained that "the extension work has overshadowed the experimental work in the public mind; as one puts it, the spectacular and publicity features of agricultural extension tend to emphasize that service and to popularize it at the expense of agricultural research." To many it appeared that the creation of an extension program had adversely affected the stations.[33]

Ironically the extension service had been created in part to alleviate the station's burden of public service, but public service,

[31]T. R. Bryant et al., "Report of the Special Committee to Study Types of Extension Organization and Policy in the Land-Grant Colleges," *AAACES*, 1914, 260–262.

[32]Davenport, "Activities under the Lever Bill" (11 September 1914, Subj. File, box 83, UIA), p. 2. See also Davenport to C. G. Hopkins, 23 May 1914, Agronomy File, box 313, UIA.

[33]E. W. Allen, "Position and Outlook of the Experiment Stations," *AAACES*, 1919, 130.

it turned out, offered several distinct advantages. First, extension workers and especially county agents were in such great demand that stations found it hard to compete in hiring agricultural graduates. The extension service paid salaries comparable to those offered by private industry, considerably higher than those paid by the station. In addition, both state and federal governments appropriated far more money to extension than to the station. In 1919 the combined appropriation for the Hatch and Adams Act totaled $1,440,000, while that for Smith-Lever was over $4,500,000 from the federal government and an equal amount from each state.[34]

Just as the station had been used to promote the college, many considered extension work a way to promote station research, which was, after all, Davenport's original intention. But as Davenport learned, convincing farmers of the value of research was far more difficult than simply equating experimentation with increased farm efficiency and profits. Indeed, such an equation could be positively counterproductive if it instilled in farmers a passion for receiving free advice without an appreciation of where and how such advice originated.

Whereas the Smith-Lever Act did benefit research to the extent that it liberated most of the station staff from routine extension tasks, research ideals of the college and station remained something of a nonissue to farmers. As F. B. Mumford pointed out, "The avenue through which favorable public sentiment has been secured in these institutions has been the extension service."[35] Regardless of whether public service was called station work or extension work, the lesson was clear. Farmers supported highly visible college functions that attempted to rationalize farm operations. The invisible functions of the college—research and education—were of considerably less interest. If they considered

[34]Allen, pp. 131–132; R. L. Watts, "Additional Federal Support for Experiment Station Work," *AAACES*, 1919, 253–255.

[35]F. B. Mumford, "Cooperation in Extension Work between the USDA and the Colleges of Agriculture," *AAACES*, 1912, 137.

these functions at all, farmers regarded them as the price one paid for the more utilitarian benefits.

Both before and after Smith-Lever, the public services of the college became substantially more visible through the network of farm advisers and farm bureaus. More than any other farm-college link, these directly joined the college and the farm community. For better or worse, this arrangement guaranteed to both farmer and agriculturalist that each would have a voice in the activities of the other. It also shaped the future client-patron relationship between the farmer and the college and helped determine the kind of responsibilities each felt toward the other. It soon became clear that while the college had labored long and hard to awaken the farmers to their own interests, farmers, once awake, knew precisely how these interests could best be served.

Farm and College: The Negotiation of Authority

The first Illinois farm bureau, in spirit if not in name, was organized in 1912 in DeKalb County. The DeKalb County Soil Improvement Association was a coalition of county farmers, bankers, and newspapermen who felt that farming could be rationalized and systematized with the help of an agricultural adviser. The board of directors included three representatives each from the Farmers' Institute, the Illinois Bankers' Association, and the Newspapermen's Association, as well as one person from each of the nineteen townships. The goals of the association were straightforward: "to promote a more permanent system of agriculture in DeKalb County; to encourage the dissemination of agricultural information in every feasible manner; and to supervise, direct, and assist in the activities of the DeKalb County agricultural adviser."[36]

Although there was no formal relationship between them, the

[36]John J. Lacey, *Farm Bureau in Illinois* (1965), pp. 13–14.

association looked to the agricultural college in selecting its adviser, W. G. Eckhardt, a protégé of Cyril G. Hopkins. The choice reflected the association's response to increasing interest among farmers in the soil-fertility issues so loudly broadcast by Hopkins. Farmers had become convinced that soil was the most significant variable in determining their profit margin; now they hoped Eckhardt would analyze their soils and recommend appropriate action. They were not disappointed. Eckhardt preached a gospel of limestone and in the process created a market for it that ultimately engendered a buying cooperative (this would later be a classic Farm Bureau pattern). Eckhardt also advised farmers on field preparation, encouraged the use of clover and alfalfa, and studied which pests and diseases were responsible for poor crops.[37]

Other Illinois counties soon followed DeKalb's initiative, and by 1914 over a dozen had hired farm advisers and created organizations to support them. In every case the organization reflected a convergence of farming, banking, and commercial interests. As one Farm Bureau official pointed out, a farmer's prosperity benefited the entire rural community, not least the local banker, whose own solvency was directly related to the farmer's success. Businesspeople's interest in Farm Bureau organization is less easily explained; while they might hope to benefit from farmers' growing sophistication as consumers, they frequently found themselves in competition with Farm Bureau cooperatives that could obtain reduced rates for farm materials by purchasing in bulk.[38]

When the Smith-Lever Act took effect in 1914, then, a pattern of farmers' organization was already in place in Illinois. Indeed, one historian has noted, "the Smith-Lever Act . . . is almost as much a charter for the Farm Bureau as it is for the extension service." College officials perceived such organizations as the natural conduit through which the agricultural expertise of the extension service would reach farmers. Rather than reinvent a more strictly

[37]Ibid., pp. 18–20.
[38]Ibid., p. 13.

educational organization, the college worked out a cooperative arrangement with the bureaus.[39]

The role of the Farm Bureau, according to Beverly Galloway, was to serve as "a headquarters or clearing house for agricultural information and for the extension work of the college . . . and other such agencies as may wish to cooperate." To effect cooperation, the bureaus were linked to the college both structurally and financially. Each Farm Bureau employed a farm adviser, who was responsible to the state extension leader, located at the college. The state leader was, in effect, an organization person for the college and helped in setting extension policy, organizing county associations, and placing advisers in the counties. The college also guided the Farm Bureau on an economic level, sharing the cost of the adviser with the USDA and the counties themselves.[40]

The farm adviser was the critical link between college and county interests and the most reliable mechanism by which the college could exercise its authority in the county. Although the appointment of farm advisers was theoretically a cooperative venture between the county and the college, in fact the counties were allowed only a token role. In 1915 counties wishing to organize a Farm Bureau were obliged to hire not a local expert but an adviser approved by the college. These advisers had to meet stringent qualifications; each had to have five years of postgraduate experience after receiving a degree from an agricultural college. Needless to say, the pool of such advisers was woefully small. Salaries were attractive—often as high as four thousand dollars a year, good by any agricultural standards—but smaller and less-affluent counties could not pay so much and therefore had difficulty attracting advisers. Some had to wait over a year before finally employing one.[41]

[39]Christiana McFadden Campbell, *The Farm Bureau and the New Deal* (1962), p. 5.

[40]Beverly T. Galloway, "Organization of Cooperative Extension Work: Machinery and Method (in the State)," *AAACES*, 1915, 224; L. A. Clinton, discussion of Galloway's "Organization," ibid., pp. 225–226.

[41]Lacey, *Farm Bureau*, p. 33.

The relationship between the county Farm Bureaus and the agricultural college was a curious accommodation of the interests of both groups. From the counties' perspective, the college was not a critical component in developing bureaus, for there was growing local initiative in both organizing farmers and securing funding. For some, in fact, affiliation with the college may have created obstacles to effective organization in that the college's standards for advisers were much stiffer than was warranted by the needs of the county. In addition, the counties may have felt somewhat overshadowed by the college, preferring to retain autonomy by relying on local funding rather than relinquish a fraction of their decision-making power to the college. On the other hand, county Farm Bureaus had much to gain by affiliation with the college. Most obviously, the Farm Bureaus could claim a degree of authority that was impossible without college patronage. By helping sponsor the bureaus the college was giving them its stamp of approval, acknowledging that the bureaus were indeed local agents and representatives of the agricultural college. By means of the farm advisers the bureaus were kept up to date on agricultural research at the college and could offer farmers fresh ideas and solutions to age-old problems.[42]

For the agricultural college, affiliation with the Farm Bureaus was also a mixed blessing. Since the organization of many Farm Bureaus preceded the Smith-Lever legislation, the college had a limited role in designing the structure of their relationship. They were faced, in a sense, with a fait accompli and were in many instances obliged to follow, rather than lead, the Farm Bureaus in matters of policy and organization. Because the burden of their authority rested almost entirely with the farm advisers, these advisers were chosen with great care.

[42]On the expanding role of the farm advisers, see G. H. Coffey to Davenport, 22 January 1921, Administrative File, box 3, UIA. The extension staff was always vigorously loyal to the Farm Bureau. For curiously defensive descriptions of their work, see E. T. Robbins, "The Farm Bureau," and D. O. Thompson, "The Illinois Agricultural Association," both in *Illinois Agricultural Policy* (1922), pp. 71–78, 79–81.

Furthermore, while the college was a strong advocate of the principle of self-help and encouraged farmers to organize themselves, striking a balance between farmers' interests and college interests was difficult. As a public institution, the college was committed to helping all types of farmers, both successful and not so successful; the Farm Bureaus, on the other hand, tended to attract a disproportionate number of progressive farmers. Some agriculturalists were concerned that the college would have little latitude in determining the extension clientele. As one extension director put it, "I can see great value in a body of men organized to promote agricultural development in the county . . . ; but I also see grave danger that this body of progressive farmers will monopolize the services of the county agent, that the great missionary work of this service will thereby be hindered, that we will not benefit those who need our services most vitally."[43]

The problem was not with the intentions of the college or the spirit of cooperation between it and the Farm Bureaus, for in those respects the arrangement seemed to work well. The problem was with the nature of field service. Once advisers were established in the county, operating out of the Farm Bureau office, driving the Farm Bureau car, using Farm Bureau stationery, to what extent would they remain loyal to the college? In effect, to what extent were the agendas of the two groups ultimately at odds with each other? As Gladys Baker, a contemporary analyst of the county agent system, observed: "Since the county farm bureaus in most counties make large contributions to the salary of the county agent, . . . it is natural that the agent feels primary responsibility to farm bureau members who contribute membership dues of $15." Under these circumstances it was difficult for the college to exercise its authority over either the advisers or the Farm Bureau, and in general, extension officials interfered little with the advisers' programs.[44]

In Illinois this undercurrent of divided interests was brought

[43]Ousley, discussion of Galloway's "Organization," *AAACES*, 1915, 226.
[44]Baker, *The County Agent*, p. 140.

into bold relief with the creation of the Illinois Agricultural Association in 1916. The IAA was organized by farm advisers who wanted to gather for both social and political reasons. Since they were all fairly isolated from the college and from one another, but shared a common set of experiences and expectations, they decided to arrange meetings at which they could compare such experiences and information. They also understood that they could enhance their effectiveness in the wider agricultural context by banding together formally. That they considered themselves lobbyists for farm interests was apparent at their first meeting in 1915. Virtually all the issues they discussed had a strong legislative component. For example, they recommended pasteurization of milk, enforcement of hog cholera laws, prohibition of the importation of infected cattle into the state, and state and federal indemnities for destruction of animals infected with foot-and-mouth disease.[45]

Under the slogan "Organized for Business," the IAA signified a break with the standard Farm Bureau approach. While the county bureaus continued to play an educative role, the IAA focused on the commercial problems faced by farmers. The IAA transformed the Farm Bureau from an appendage of the extension service into a semiautonomous advocate of the farmer's position on economic matters. The stated goals of the IAA in 1916 leave no question as to the nature of its interests and activities. In addition to sponsoring legislation, it aimed to work out a more efficient marketing system, encouraged livestock production, published information on rural betterment, and bought farm materials cooperatively.[46]

It would be a mistake, however, to treat the Farm Bureau and the IAA as entirely separate; indeed, contemporary writers as well as historians have had difficulty determining where the Farm Bureau stopped and the IAA began. Some writers treat the two as identical; some consider them distinct in their policies. This ambiguity was deliberate. Whereas the Farm Bureau was

[45]Lacey, *Farm Bureau*, p. 42.

[46]Campbell, p. 7; True, *History of Agricultural Extension*, pp. 155–156.

funded as an educational agency, and farm advisers were forbidden to engage in any commercial activity, the IAA was aggressively commercial. Yet they occupied the same offices, shared the same staff and clientele, and in point of fact, farm advisers were members of the IAA. It is no wonder that the IAA had a virtual monopoly on extension activity in Illinois; rather, the wonder is that the university trustees could accommodate this relationship as long as they did.[47]

This tension between the role of the Farm Bureau and the role of the college was personified in the farm advisers, who in trying to serve two masters incurred the displeasure of everyone. College officials insisted that they restrict their activities to education, but the line between educational and commercial advice was slim indeed. Especially in the 1930s, advisers were under attack from businesspeople, who were quick to note a farm adviser's signature on a cooperative marketing order. Local businesspeople tended to feel, not surprisingly, that advisers were running them out of business by choosing suppliers who delivered less-expensive or higher-quality goods, rather than boosting local entrepreneurs. In the 1920s during the hog cholera epidemic, advisers who demonstrated vaccination techniques to isolated farmers came under vehement attack for trying to drive veterinarians out of business.[48]

[47]Campbell, p. 6. Largely because of this ambiguity, the university severed its relationship with the Farm Bureau in 1937. See William J. Block, *The Separation of the Farm Bureau from the Extension Service* (1960). As late as 1933 officials continued to defend the arrangement, contending that the Farm Bureau and the extension service were separate but equal. See George E. Farrell to W. M. Stickney, 23 February 1933, 8/1/2, box 17, file 30, UIA.

[48]W. F. Handschin, "The County Agent and the Farm Bureau," *AAACES* (1919), 287–290. In 1917 Bradford Knapp warned against the temptation among farm advisers to transact business for the bureau; see his "Relationship of Agricultural Extension Work to Farmers' Cooperative Buying and Selling Organizations," *AAACES*, 1917, 305–310. For a critique of the favoritism displayed by farm advisers to Farm Bureau members, see C. B. Smith to H. W. Mumford, 20 September 1932; for advisers' recommendations to farmers to join the Illinois Grain Corporation, see C. E. Johnson to USDA Extension, 12 January 1933; for advisers' role in marketing, see Leonard Braham to J. C. Spitler, 30 January 1933; and on

Despite these difficulties, faculty members in the agricultural college were generally in favor of the IAA. In 1924 Herbert W. Mumford, dean of the agricultural college after Davenport's retirement in 1922, asked his colleagues for their thoughts on the proper role of the IAA in relation to both the college and the farmers of Illinois. Only one of the dozen or so respondents criticized the IAA for either its commercial activities or its dominant role in Farm Bureau matters. Most felt that the proper sphere of activity for the IAA centered on legislation, taxation, and cooperative marketing issues. Agronomist W. L. Burlison went so far as to suggest that the IAA should represent farmers as an "industrial or business body to other social or economic units" and should maintain "broad contact with all lines of industry which in any way has [*sic*] to do with agriculture." In a speech before the IAA, Mumford described the interests of the college and the Farm Bureau as "mutual" in that both organizations shared the goal of "making farming a profitable business."[49]

By 1920 this sentiment was no more than a reflection of reality. The postwar depression dealt farmers a serious blow; in raw numbers farm prices dropped more sharply in 1920 than they did in 1930.[50] Farmers were extremely concerned about economic issues that had been nonexistent less than ten years before. Increasingly the college shifted its attention from production to marketing and economics, attempting to restore the balance among research, efficiency, and profit.

---

vaccination, see Fred Henke to Spitler, 30 January 1933, all in 8/1/2, box 17, file 30, UIA. This file contains voluminous correspondence and reprints relating to the confusing, ambiguous role of the advisers.

[49]H. W. Mumford to J. C. Spitler (extension), F. A. Gougler (farm adviser), J. D. Bilsborrow (state leader, extension), Fred Rankin (extension), W. H. Smith (state leader, extension), H. M. Case (Department of Farm Management), E. T. Robbins (livestock extension), Wheeler Bull (Department of Animal Husbandry), E. H. Lehmann (Department of Farm Mechanics), W. L. Burlison (Department of Agronomy), 14 February 1924, and their replies, and Mumford, "The College of Agriculture and the Farm Bureau" (address at the 8th Annual Meeting of the IAA, 1924), p. 10, all in 8/1/2, box 25, UIA.

[50]*Historical Statistics of the United States* (Washington, D.C.: Bureau of the Census, 1976), p. 511.

The Failure of the Design

In 1909 the Carnegie Foundation issued its fourth annual report on American education, and it was marked by a strident attack on the land-grant colleges. The foundation was disturbed to find that in terms of curriculum, entrance requirements, and pedagogic sophistication, these colleges resembled secondary schools rather than universities. They blamed the ongoing struggle for existence, which forced these colleges to concentrate on filling the rolls with students rather than developing a rational and intellectually rigorous course of instruction. As a result the colleges were overcrowded and lacked a clear mission; the foundation was unable to determine what particular role the land-grant college was in fact playing. It neither taught students to become better farmers nor provided adult farmers with innovative advice. Neither trade school nor liberal arts college, the land-grant school had failed to convince the Carnegie Foundation that agricultural science would define institutional goals.[51]

While the Carnegie report may have seemed unduly harsh to land-grant administrators, there was a measure of truth in its findings. Some specific indictments could be brushed aside as trivial or wrong, but on the whole they presented an image of the colleges that was disappointingly accurate. The agriculturalists were proud of their considerable achievements, especially from 1900 to 1910, but these could not negate the essentially ad hoc nature of agricultural development in land-grant institutions. Like federal agricultural policy itself, land-grant policy and programs were created in response to crises. When Davenport arrived at Illinois he reacted to the crisis in facilities and student population by selling farmers on the experiment station. When the station was subsequently overwhelmed with farmers' demands, Davenport created an extension service, which, within a few years, was understaffed, underfunded, and in competition with both the college and the station.

[51]Carnegie Foundation for the Advancement of Teaching, *Fourth Annual Report* (1909), pp. 83–111.

Once Smith-Lever was enacted, the machine was in place, but it was neither well oiled nor finely tuned. The three divisions continued to quarrel and complain. In the late 1910s and early 1920s the stations were called the weakest link in the agricultural programs, and station directors agitated again for more federal funding, which, they hoped, would increase the distance between station and extension agendas. Station directors were disgruntled that researchers were often required to teach in the college, thus diminishing station autonomy.

The agricultural colleges developed a diverse agenda for research, education, and extension and then found themselves in many ways unable to juggle these three responsibilities. In the early years the functional division of agriculture into research, education, and extension components seemed unproblematic, but by 1920 it was clear that agricultural knowledge itself was not a unified, coherent subject. Not all research was appropriate for extension and not all farmers' needs and interests were worthy of research. Yet the machinery was designed as though research, education, and extension were compatible, or as though research were a pool from which extension and education naturally and inevitably drew their resources.

If Davenport left a legacy at the University of Illinois, it was in establishing a reciprocal relationship between the college and the farm community. Farmers during and after his tenure seemed almost instinctively inclined both to rely on the college and to direct its course. They took Davenport's philosophy almost too seriously and were often more in charge of running the machine than the college's own faculty and administrators.

The relationship between farm and college was ambiguous and the role of the college unclear in the wake of Davenport's departure. The query of one observer describes the quandary in which agriculturalists found themselves: "Should the agricultural colleges created by public act teach what the public wants or what the public ought to be taught?"[52] The college had become neither a master nor a servant, neither a brave new world nor a defender

[52]W. J. Kerr, "Some Land-Grant College Problems," *AAACES*, 1911, 37–51.

of the faith, but something in between. It had become a negotiator both within and outside its walls, attempting to mediate internal disputes over funding, divisional autonomy, and policy and external disputes over innovations, economic difficulties, crop surpluses, and marketing on a regional and national scale. As we examine their corn development program, we shall see how Davenport's machine—originally intended as a platform for sophisticated research—instead provided an ambivalent home to both farming and scientific interests.

4

High Science, Low Science:

Research and Extension at the University of Illinois, 1910–1935

As Eugene Davenport prepared to retire after twenty-seven years as dean, he paused to reflect on the growth and success of the College of Agriculture and to consider what the future might hold. Writing with a mixture of nostalgia and vision to W. L. Burlison, head of the Agronomy Department, Davenport invoked his long experience as research entrepreneur and farmers' advocate. In setting new policies, Davenport suggested, it was not sufficient to concentrate on persuading all farmers to adopt new practices and recommendations; indeed, such a strategy would only alienate the more progressive farmers, who were eager to learn the newest techniques and who, incidentally, formed the backbone of the college's support system. Rather, the department should constantly tackle new problems and create new research programs that generated questions as well as answers. Moreover, to Davenport the direction research should take was clear: he predicted that the next era of agricultural research would focus on "physiological questions of crop production." As an animal breeder, a member of the American Breeders' Association, and a vigorous participant in the Association of American Agricultural Colleges and Experiment Stations, Davenport had become convinced that genetics and physiology together would provide the tools for studying critical problems of agricultural production.[1]

[1]Eugene Davenport to W. L. Burlison, 23 November 1921, W. L. Burlison File, 1920–1922, folder 312, UIA.

Between 1910 and 1935 these two sciences did indeed form the basis of crop investigations at Illinois. Under an expanding and increasingly sophisticated scientific staff, crop research came to reflect an aggressive attempt to apply theoretical science to the chronic and mundane problems of crop production. While the physiologists sought to describe the behavior of plants, geneticists tried to understand the patterns of inheritance. The two approaches worked symbiotically: breeding techniques helped physiologists isolate particular plant characteristics; physiological studies of form and function gave geneticists a map from which heritable characters could be derived and identified.

But by 1921 research was but one component of the crops program. In that year the Division of Crops Extension was established and, with it, a rather different approach to farm problems. Whereas the motto for researchers might have been "innovation," for extension specialists it was more realistically "evolution," with all the painful slowness the term implied. And with respect to corn, the goals of research and extension differed markedly. Although this situation did not correlate with college rhetoric that emphasized a natural consistency between research and extension, it nevertheless was an apt reflection of the very different concerns of the two divisions. Research responded to larger scientific trends, and extension was restricted to what were essentially market forces; extension specialists taught what farmers would listen to, sprinkling their lessons with as much new information as the clientele was ready to assimilate.

For the next fifteen years, the major concern of Crops Extension was educating farmers to field-select their seed corn according to guidelines established by the college. This program combined traditional crop improvement practices familiar to farmers with new practices resulting from recent investigations of corn disease. In an effort to make crop production more efficient, rewarding, and scientific, Crops Extension sought to provide farmers with the knowledge they needed to accommodate such research findings. A corollary project of Crops Extension, geared to seed producers rather than ordinary corn farmers, was the popularization of seed certification. Together the two projects represented the college's attempt to rationalize corn production in Illinois by coordinating the interests of farmers, growers, and the college.

Breeding at Illinois, 1910–1920

Following the exodus of Holden, Love, and East from Illinois by 1905, the corn-breeding work was left where it had begun—in the hands of L. H. Smith, Hopkins's first assistant. For the next fifteen years, Smith worked virtually alone to develop and implement corn projects at the Illinois station. He not only continued work already under way but devised new approaches to crop improvement and basic research. Along traditional lines he continued Hopkins's oil and protein selection experiments and revived the comparative field trials of corn, in which a number of different varieties were grown side by side in comparable fields around the state to determine which varieties were best adapted to particular corn-growing regions. This information enabled the station to help farmers across the state select appropriate strains.[2]

Smith also extended Hopkins's ear-to-row selection method beyond the chemical studies, applying it to gross physical characteristics of corn. In 1903 he began selecting corn plants that carried the ear either high or low on the stalk. His aim was to determine if simple selection could in fact change the position of the ear, just as it had changed the kernel's chemical composition. After ten years of selection, Smith found that he had altered the ear position by four and a half feet overall, a striking affirmation of the power of selection for particular characteristics. This experiment also had a more practical rationale, being in part an investigation of the extent to which mechanical corn pickers could be designed to harvest corn more efficiently than farmers themselves.[3]

[2]Agronomy Department Projects, n.d. [post-1926], 8/2/0/4/, box 1, UIA, p. 31.

[3]Office of Experiment Stations (OES), *Report*, 1911, p. 44; L. H. Smith, "The Effect of Selection on Certain Physical Characters in the Corn Plant," *Illinois AES Bulletin* 132, 1909.

In 1910, Smith turned his attention more explicitly to breeding for improvement using Mendelian methods. One might surmise that his breeding efforts were informed by both the 1906 lecture tour in which Hugo De Vries discussed small grain breeding in Europe on stops in Chicago and Bloomington, Illinois, and the investigations of East and Shull from 1907 to 1909. Smith's work in this period used methodology and design similar to theirs and, not surprisingly, resembled too the more eclectic breeding investigations represented by the American Breeders' Association.

In the next few years Smith divided his efforts between corn and small grains. He developed two new wheat strains by selecting variations from popular lines and inbreeding them for several generations.[4] He also tried his hand at soybean breeding and in 1910 made hundreds of selections and tests. To test Johannsen's pure-line theory, which held that an inbred plant was "pure" for a particular characteristic and would breed true, Smith attempted to alter the characteristic in a pure line of soybeans by means of selection. He had, of course, been successful in altering characteristics in heterogeneous, open-pollinated corn plants. But with inbred soybeans he had no luck; as predicted, selection had no effect whatsoever on pure-line strains.[5]

Between 1911 and 1915 Smith undertook corn studies that in several respects marked a break with traditional corn work at Illinois. In embracing the new Mendelian methods for both understanding corn inheritance and improving the corn crop, Smith rejected routine methods of improvement because, he said, they did not work. First, in 1912 Smith began a ten-year experiment to test the ear-to-row selection method yet again. This time Smith

[4]Wheat is naturally self-fertilized and therefore in some respects easier to breed than corn. The first strain, called Ilred, was selected from Turkey Red wheat in 1910 and distributed to farmers in 1921; the second, Illinois number 2, was selected from Indiana Swamp wheat in 1915 and distributed in 1931. Illinois number 3 was of further use in the mid-1930s, when it was selected from to develop a strain called Prairie. University of Illinois, Department of Agronomy, "Historical Data for President Kinley," February 1941 (8/6/2, UIA), pp. 28–29, 32–34.

[5]Smith did not publish the results of these soybean experiments.

wanted to determine whether selection could increase the yield of corn, by far the most important overall characteristic from a practical standpoint. Although the method was successful in altering other characters, Smith concluded that it did not increase yield and in fact was no better than selection as a method of improvement. Next, in a 1914 breeding experiment, Smith compared the yield of different crosses of open-pollinated varieties and reported that "no benefit was observed from crossing commercial [open-pollinated] varieties."[6]

The Mendelian studies at Illinois, as at other experiment stations, were investigative rather than strictly practical in the mid-1910s, although many were conducted with the long-term goal of crop improvement. The corn-breeding work was, in fact, more theoretical than that for other crops. Reporting to the Office of Experiment Stations in 1911, Smith explained that corn breeding "was taken up from the standpoint of unit characters and Mendelian transmission. Similar work was done with other field crops but mainly with a view to improvement." Smith may have been referring here to the small-grain breeding, as well as the Mendelian study of inheritance in peas, apples, and strawberries.[7]

Smith's corn investigations were characteristic of the more general Mendelian studies undertaken elsewhere on corn and other plants and animals. Such studies began by studying the inheritance of particular characteristics as well as the correlation of such characters between parent and offspring. For instance, if the corn ear had ten rows of kernels, the number of offspring that also had ten rows would be compared with the number that did not, in an attempt to derive or verify Mendel's predictive rules of

[6]OES, *Report* (1914), 34. L. H. Smith and A. M. Brunson, "Experiments in Crossing Varieties as a Means of Improving Productiveness in Corn," *Illinois AES Bulletin 306* (1928). Noting that crossing varieties did not seem to result in higher-yielding corn, Smith stated that "these experiments . . . do not deal with that more complex, though highly promising, plan of corn improvement which involves the production and the subsequent crossing of self-fertilized lines."

[7]OES, *Report*, 1911, 104.

inheritance. By 1914 Smith could report to Davenport that this "fundamental" work "is proving to be most illuminating regarding the habits and nature of the corn plant."[8]

These Mendelian studies also extended to animals, and in 1915 and 1916 animal breeders at Illinois began working on virtually the same questions that guided plant breeders. To determine whether inbreeding caused a loss of vigor, as had been observed in corn, breeders cross-fertilized pig siblings (the closest thing to inbreeding in animals). Pigs were also used to investigate whether particular characteristics were inherited singly or together, and in 1916 the same questions were explored by means of four hundred sibling-bred mice. On a more commercial level, breeders launched a project on breeding explicitly for particular characters, using skunks "to learn how to breed for a black coat which shall be long and also silky." As an unanticipated bonus the breeders by 1915 had acquired two "mutants," a pure white skunk and a black-eyed skunk.[9]

Between 1917 and 1924 the breeding program at the Illinois station was expanded with the appointment of four men. George H. Dungan was appointed in crop production in 1917 after receiving his B.S. at Illinois. Three years later he finished his master's and went on to receive his Ph.D. in plant pathology and physiology at Wisconsin. In 1920 C. M. Woodworth, who had received a Ph.D. in plant genetics from Wisconsin in that year, was appointed professor of plant breeding and genetics at the Illinois station. F. L. Winter was named assistant station agronomist in 1922 after receiving his B.S. at Illinois. He subsequently received his M.S. there and in 1927 his Ph.D. in plant breeding and genetics from Cornell. Benjamin Koehler was named crop pathologist in 1924 and received his Ph.D. from Wisconsin the following year. All but Winter stayed with the university for the remainder of their professional lives.[10]

[8]Ibid.; L. H. Smith to Davenport, 27 October 1914, file 313, UIA.

[9]OES, *Report*, 1915, 107; 1916, 107.

[10]After seven years at Illinois, Winter became consulting agronomist with the Hoopeston Canning Company in Illinois and in 1933 joined the Associated Seed

During this seven-year period the corn-breeding program underwent a transformation. Although Smith had considerably enlarged the breeding program to include genetics and had extended the use of genetics to other crops, he could not long remain the sole authority on breeding. By the early 1920s a sizable number of professional plant geneticists came of age, the first generation of breeders who had been systematically trained to use Mendelian methods and explanations. Indeed, one might consider this rapid increase in plant-breeding positions at Illinois as Davenport's swan song; three of the four appointments were made before his retirement in 1922, and the fourth, Koehler, was cooperating with the station from his position at Funk Brothers Seed Company before he was officially hired. Further, Davenport's commitment to a strong and professionally respectable program was reflected in the training these new scientists had received. All had obtained doctorates from America's leading programs in agricultural genetics: Wisconsin and Cornell were without question two of the three most sophisticated schools applying Mendelian methods to agriculture. By committing the station to this agricultural innovation, Davenport was assured that his own research would be continued after his departure.[11]

Research in the 1920s

In a 1921 report on conditions at the experiment station, W. L. Burlison attempted to summarize the results of station investigations and to suggest directions for further research. Although corn investigations were mentioned only briefly in this

Growers as resident geneticist. All biographical information is from *American Men and Women of Science*, 9th ed.

[11]See Barbara A. Kimmelman, "A Progressive Era Discipline: Genetics at American Agricultural Colleges and Experiment Stations, 1890–1920" (Ph.D. diss., University of Pennsylvania, 1987). For reasons that remain unclear, Smith left in 1922 to become chief in charge of publications of the soil survey at the station.

overview, the overarching research program for corn, as for other grains, was fairly distinct. Following Davenport's lead, Burlison created an agenda in 1921 that guided the program in plant investigations for many years.[12]

Burlison outlined a crop improvement program in which corn, as well as other plants, would be approached from two directions. Physiological investigations would focus on "learning the life history of each of our common field crops" and determining the relationship between plants and the field environment. It would also include testing varieties for adaptation to actual farming conditions and acquiring "knowledge of the principles involved in variety adaptation." Studies in plant inheritance, on the other hand, would use breeding techniques to learn "the methods of hereditary transmission of plant characters." "This knowledge," Burlison continued, "should be applied in increasing the yield and improving the quality of our several farm crops." Whether corn was studied from a physiological or hereditarian point of view, then, the object was the same: to discover general laws of plant behavior and apply these laws to crop improvement.[13]

Burlison's report the following year to the Soils and Crops Advisory Board echoed these approaches and reemphasized the importance of applying theoretical principles to improvement. Studies of corn inheritance focused on those characters that influenced yield, such as disease resistance, and the correlation of characters in parent and offspring was considered through the lens of disease immunity. Physiological studies centered on establishing the relative importance of many environmental and physiological factors and the relations between them in their

[12]The predominant concern of the station was soil investigations, for which over three-fourths of the staff members were hired. The range of station projects reflected in this report also included apple diseases, cattle breeding, cost accounting, sheep feeding, a comparison of the relative merits of horses and tractors, and forage crop research. "Conditions in the Agricultural Experiment Station of the University of Illinois," January 1921, 8/1/2, box 19, UIA.

[13]Ibid.

endless variations and combinations. Emphasizing that "the basic function of the Experiment Station is to conduct research," Burlison nonetheless anchored his projects firmly to the cause of crop improvement.[14]

These two approaches to crops were reflected in the research projects undertaken by the station staff in the 1920s. George Dungan led physiological investigations that examined the effects on yield of harvesting at different stages in the corn plant's maturity (1920) and planting at different times (1924). These were without doubt the sort of studies that appealed to farmers, who worried about the losses in yield they would suffer when late spring rains delayed planting or an early thaw tempted them to plant the crop earlier than usual.

The Mendelian studies, though of less immediate use to farmers, also promised more spectacular results. By 1921, when the station began to study corn inheritance systematically, Burlison had decided that the virtues of selection were limited. "From the experiments conducted at this station," Burlison reported, "it appears that while selection has had some effect in increasing the yield of corn, it has not given the results that were anticipated. It is proposed to attack this problem of breeding for yield from a slightly different angle." This new "angle" was inbreeding and crossing corn to produce single-cross hybrids.[15]

Of course inbreeding was not new, either to the plant breeders or to the Illinois station. What was new was the station's self-conscious and systematic investment in genetics as a method of permanent plant improvement. Inbreeding and crossing corn in this scheme was but one of three modes of Mendelian investigation at the station. Theoretical studies were also begun: in the mid 1920s W. B. Gernert reportedly carried on extensive, albeit unpublished, studies in corn inheritance; W. J. Mumm studied ge-

[14]W. L. Burlison, "The Present Status of the Work of the Agronomy Department and a Program for Future Development," 18 January 1922, 8/1/2, box 19, UIA, p. 8.

[15]W. L. Burlison, "Report to the Advisory Committee" (January 1921, Alexander Collection).

netic linkage and mutations in corn; and C. M. Woodworth identified particular heritable characters.[16]

The theoretical corn genetics work paralleled research on other crops at the station. In 1920 the genetic study of soybeans became a particularly compelling research topic, not only because soybeans were a relatively new crop, which agronomists favored in a system of crop rotation, but also because the soybean seemed an appropriate candidate for "directional" breeding, in this case for high oil content. Oil content was the most economically salient character of soybeans, and the desire to increase it by means of inbreeding and crossing was irresistible. Oats and barley were also included in these early genetic studies. As with corn, the object was to discover "the mode of inheritance of morphological and physiological characters," as well as to isolate and develop "economically desirable strains."[17]

The station's third type of breeding work was a cooperative project with J. R. Holbert, a USDA plant breeder at Funk Brothers Seed Company. In 1917 the USDA had established a federal field station at Funk Brothers, with Holbert in charge, for the purpose of studying corn diseases. In 1919 Holbert hired Benjamin Koehler, who was then earning his Ph.D. in plant pathology at Wisconsin, to assist him in this work, and they began cooperating unofficially with the Agronomy Department at Illinois. This project, which continued into the 1930s, was a curious blend of old and new methods of corn breeding, employing both selection and inbreeding as part of a comprehensive strategy to eliminate prevalent corn diseases. The work also formed the basis of subsequent extension recommendations to farmers growing corn.[18]

[16]"Historical Data for President Kinley," p. 28; C. M. Woodworth was the most theoretically inclined; see, e.g., his "Heritable Characters of Maize: 28-Barren Sterile," *Journal of Heredity*, 1926, 17:405–411; W. J. Mumm and C. M. Woodworth, "Heritable Characters of Maize: 36-A Factor for Soft Starch in Dent Corn," *Journal of Heredity*, 1930, 21:503–505.

[17]Agronomy Department Projects, pp. 18, 30; "Historical Data for President Kinley," p. 30; Agronomy Department Report, 1921, 8/2/0/4, box 1, UIA, p. 29.

[18]The roles of Holbert and Funk Brothers in the disease project are recounted in Cavenagh, *Seed, Soil, and Science*, pp. 216–231.

In the first phase of the "corn root and stalk rot disease project," researchers compared disease resistance in ordinary open-pollinated corn with that in inbred lines by inoculating the corn plants with the disease-producing organisms in both the field and greenhouse. Once the differences in resistance were detected, they attempted to correlate this tendency of some plants to resist the disease with those plants' morphological and physiological characters. The object of the correlation study was to develop a code for visually identifying those plants that were most susceptible and most resistant to the disease. The final phase of the work consisted of identifying, maintaining, and ultimately crossing those inbreds that demonstrated superior resistance.[19]

Holbert, Koehler, and Dungan concluded fairly early in the project that some visible features of corn consistently correlated with root and stalk rot. By 1922 they had established five demonstration plots throughout the state to illustrate these differences, and apparently, corn growers found the results striking. According to J. C. Hackleman, crops extension specialist, these demonstration plots created a demand from farmers for a corn show that would "organize some of the points which experimental data were beginning to show were distinguishable."[20]

In many respects this corn-disease project epitomized the dual role the college attempted to fill, both conducting long-term research and providing farmers with meaningful, up-to-date advice on crop production. The station was aware of the necessity for devising research projects that could be extended to farmers in a serial or piecemeal fashion; in the corn-disease project this neces-

[19]"Historical Data for President Kinley," pp. 20–21; Agronomy Department Projects, p. 24 and passim. The results of this project were codified in J. R. Holbert, W. L. Burlison, B. Koehler, C. M. Woodworth, and G. Dungan, "Corn Root, Stalk, and Ear Rot Diseases and Their Control through Selection and Breeding," *Illinois AES Bulletin* 255, 1924. See also F. L. Winter, "The Effectiveness of Seed Corn Selection Based on Ear Characters," *Journal of the American Society of Agronomy*, 1925, 17:113–118.

[20]J. C. Hackleman, "Annual Report of the Crops Extension Specialist for 1922," p. 3. These reports are kept in the office of the current crops extension specialist, University of Illinois, Department of Agronomy.

sity was met. Scientists could explore the theoretical possibilities of breeding for disease resistance, while providing farmers with practical advice on preventing the spread of disease, and both approaches were perfectly consonant with the larger objective of disease eradication. This principle of complementarity was important because it allowed the two divisions to synchronize their efforts, presenting the farmer with an image of the various college divisions working harmoniously and coherently. Unfortunately, as with much other research, such coherence was not easily attained.

More often than not, both the nature of agricultural research and farmers themselves effectively prevented a "fit" between research and extension. Crop research consisted in large part of field, rather than laboratory, experiments, and because Illinois agronomists could conduct only one batch of field experiments per year, most projects were in the fields for a decade. In other words, the lag time between the beginning of a project and the "extension" of results to farmers was at least ten years, especially with breeding experiments, in which each generation built on the one before. Moreover, farmers themselves were notoriously slow to adopt new recommendations, and agronomists estimated another seven to ten years would pass before the majority of farmers had changed their farm practices in accordance with new guidelines.[21]

[21]For a classic study of the rate at which farmers adopted hybrid corn and the factors that influenced the timing of adoption, see Neal C. Gross, "The Diffusion of a Culture Trait in Two Iowa Townships" (Master's thesis, Iowa State University, 1942); see also Bryce Ryan and Neal Gross, "The Diffusion of Hybrid Seed Corn in Two Iowa Communities," *Rural Sociology*, 1943, 8:15–24. Discussing the rate at which farmers switched from rough to smooth corn, Hackleman lamented: "We recognize that it takes from five to ten years of work on an organized project before it is possible to notice much change in an acceptable standard or custom, even though the new has proved itself superior in practically all cases. This apparent inertia is probably due to the fact that most of our teaching must be done by the indirect method. It must be recognized that only through observation of practice on the farms of project leaders and other good farmers will the rank and file hear of or see the better practices in vogue." Hackleman, "Annual Report, 1929," pp. 6–7.

Extension in the 1920s

Seven years after Smith-Lever was enacted, the Division of Crops Extension was established at Illinois under the direction of J. C. Hackleman. Hackleman, who had obtained his B.S. in agronomy at Purdue (1910) and M.S. at Missouri (1912), came to the University of Illinois in 1917 as associate professor of crops extension after serving as assistant agronomist at the University of Missouri for seven years. In 1919 he became director of Farm Crops Extension, a position he held for the remainder of our period. Hackleman belonged to no specialist organizations and apparently did not conduct research; he was an extension worker extraordinaire, devoting his entire life to organizing and serving farm groups as well as forcefully guiding their interests. He was moved to express his evangelical commitment to extension in 1926 when reporting on the success of using local project leaders to ensure that scattered local field projects would be attended to in his absence:

> It is through the training and use of project leaders that these men begin to see the effects of their work and enjoy the satisfaction of feeling a certain personal responsibility for at least part of their committee's advancement. In short the project, be it what it may, becomes merely an instrument in the hands of the specialist and the county agent with which to single out and develop real agricultural leaders. The ultimate goal, after all, is not merely the making of more pounds of pork, more bushels of corn, or fewer and larger apples or peaches, but is in reality the making of men.[22]

The Division of Crops Extension epitomized national extension ideology and goals, and Hackleman's stated aims perfectly echoed A. C. True's ideas on what extension should be. Hackle-

[22]Hackleman, "Annual Report, 1926," p. 6; *American Men and Women of Science*, 9th ed., p. 450.

man's goals in 1921 were, first, to "carry the results of the most recent investigational work into the state" and thereby to bring farm practice into line with agricultural science; second, to "assist in standardizing crop teachings and crop demonstrations work in the United States" and so to eliminate "erroneous conclusions" in the farmers' minds; third, to work with county farm advisers in developing a statewide network of experimental demonstrations and projects through which farmers could observe and participate in newer approaches to crop production; and fourth, to help identify farming problems that required the attention of college scientists. Hackleman's understanding of the role of extension was completely consonant with the role projected in 1914, and as an abstraction, it depicted an image of perfect harmony in relations among the county, state, and federal groups, as well as between extension and research. It had the added advantage of being completely unexceptionable from the farmers' point of view.[23]

Hackleman in the 1920s focused on three projects: persuading farmers to grow more legumes, especially soybeans; teaching farmers how to field-select seed corn following certain guidelines; and popularizing a program of seed certification. The soybean campaign was part of a larger project of soil improvement in Illinois, through which the college tried to persuade farmers to conserve and improve their soil fertility by systematically planting part of their land with soybeans. Soybeans, which were introduced in the United States shortly after 1900, had proven almost magical in their ability to fix nitrogen in the soil and were therefore extremely valuable in a system of crop rotation. The station was busily breeding soybeans to produce strains adapted to Illinois soil and environmental conditions. Finding resistance among farmers to adopting soybeans as a green manure, however, Hackleman reasoned that if a market could be developed for the beans, then farmers might be more interested in trying to grow them.

Hackleman's efforts to interest both farmers and manufacturers

[23]Hackleman, "Annual Report, 1922," p. 2.

in soybean production provides a useful view into the complexities of trying to organize agriculture. In 1922 Hackleman and farm advisers contracted with four processing companies who agreed to purchase 250,000 tons (or 500,000 acres) of soybeans the following year. News of this guaranteed outlet encouraged many more farmers to plant soybeans, increasing the overall acreage in Illinois. In the fall, when it appeared that the price for beans would drop because of the large harvest, most farmers decided to store their beans for seed or to wait for higher prices. The result was that the companies received only a few thousand tons of beans. Yet even though the college may have been chagrined and embarrassed in not meeting the quota, Hackleman's logic was basically sound: farmers could be coaxed into adopting new crops and practices if given a proper economic incentive. Illinois farmers continued to alternate their corn crops with soybeans.[24]

The primary focus of crops extension in Illinois, however, was educating farmers in selecting disease-resistant seed corn. By 1921 the disease project had demonstrated sufficient correlation between corn disease and ear characteristics that Hackleman began taking the results into the farm communities. These correlations were first codified when Hackleman devised a new scorecard for corn judging that redefined standards used in identifying healthy, high-yielding ears. While nearly half the percentage points on the scorecard were reserved for "general appearance," perhaps in deference to those who could not resist the old "fancy" corn that won honors on purely aesthetic grounds, 45 percentage points were concerned with characteristics especially indicative of disease resistance. For example, 5 percent of the score was for indentation; judges were instructed to discriminate against roughly indented kernels, an indicator of slow maturing because of disease; 5 percent went to the chemical composition of the kernel, with discrimination against starchy kernels because they were more susceptible to disease. Disease was also indicated by ear butt and shank attachments (where the ear hooked onto the

[24]Ibid., p. 11.

stalk) that were pink, brown, or cracked (10 percent), and so forth.[25]

For Hackleman these new findings provided invaluable ammunition in the war against indifferent farm practices. Throughout the decade he concentrated on emphasizing to farmers the importance of using research results to increase the yield and quality of their corn. In previous years farmers had been encouraged to inspect their fields before pollination and remove cornstalks that appeared diseased or barren so that such corn would not perpetuate itself. In the 1920s this strategy was substantially improved. With the new correlation rules, farmers could remove apparently healthy corn that nevertheless displayed signs of incipient disease to the trained eye.

In 1923 Hackleman initiated a project aptly called "Better Seed Corn," through which he trained farm advisers to identify corn disease in the field; the advisers then passed the lesson on to the farmers themselves. Through an elaborate series of meetings, classes, and demonstrations, as well as a rural media blitz, Hackleman sought to generate interest in the new methods by instilling enthusiasm in his corps of advisers and in the farmers, who, he hoped, would pester the advisers for more information. The main incentive for farmers, naturally, was higher yields. Corn yields after selection tended to average five to ten bushels an acre more than fields planted with unselected seed, although the amount varied from year to year. In 1927, for example, corn disease in Illinois was especially prevalent and decreased yields by twelve to fifteen bushels an acre. Those farmers who practiced selection, however, suffered less than those who did not. Hackleman rather optimistically concluded that farmers selecting against disease lost on average only two to five bushels an acre.[26]

As with so many other extension recommendations, it is difficult to measure the results of Hackleman's educational efforts. The Soils and Crops Advisory Board applauded his strategies, pointing out that the simple physical examination of corn for

[25] Ibid., pp. 3–4.

[26] Ibid., p. 5; Hackleman, "Annual Report, 1924," pp. 19–22; "1925," pp. 6–7; "1927," p. 4; "Historical Data for President Kinley," p. 1.

disease was probably the least taxing method of corn improvement for farmers, who, they felt, would reject anything more complex and time-consuming. Nonetheless the practice did require farmers to spend considerable time in the field inspecting their corn and disposing of inferior ears, and many were unable or unwilling to tackle the work. The incomplete success of the project was apparent to Hackleman by 1929. Frustrated and pessimistic, Hackleman blamed apathy among farmers aware of the facts of agricultural life: "The toll of poor seed and of corn disease in Illinois corn fields amounts to between 70 and 80 million dollars annually. The fact that this loss is perennial, that the farmer has never seen and will probably never see a one-hundred percent disease-free field, there is considerably [*sic*] apathy and it is therefore difficult to awaken the interest that the problem really deserves."[27]

Hackleman's other project for improving farming practice was the development of seed germinators that the farmer could easily construct and operate on the farm. Germination tests were used to determine the overall viability of seeds; with the germinators, individual farmers could germinate a few seeds from a corn ear to determine if that ear was a healthy candidate for planting. Although the test was not foolproof, and indeed much diseased seed would germinate with no visible sign of disease, it was considered a corollary to selection in eliminating as much diseased seed as possible. In 1925 Hackleman "invented" a farm germinator in cooperation with the Department of Farm Mechanics and began sending plans to any farmer who requested it; yet in 1928 only eighty-five plans were sent out.[28]

Hackleman also thoroughly indoctrinated project leaders in the use of germinators. In 1925 he initiated a "corn germination school" at the college, a week-long session in which project lead-

[27]Hackleman, "Annual Report, 1929," pp. 5–6; Charles Rowe, member of the Farm Crops Advisory Committee, meeting, January 1924, 8/1/2, box 17, UIA, p. 25.

[28]Hackleman, "Annual Report, 1925," p. 10; "1928," p. 5; "Farmers Learn Seed Testing," *Prairie Farmer*, 12 February 1931, 18; "Don't Guess about Seed Corn," *Prairie Farmer*, 3 February 1930, 16–17.

ers throughout the state were taught how to build, operate, and interpret the results of the germinators. Upon returning to their communities, the project leaders then provided germination tests for farmers wanting them. Unfortunately, this work put the extension service in direct competition with commercial laboratories offering the same service. Hackleman argued that extension tests were more reliable than commercial tests because the extension service had no reason to mislead the farmer, but the commercial labs, which were often operated by the seed companies, had a vested interest in test results: "The commercial lab would hardly report to a farmer that 85% of his corn was badly diseased and therefore unfit for seed because that farmer would probably not be willing to send corn the next year. The project leader, reading his own germinator, could draw such conclusions and proceed to get more corn in test or procure better seed from some other source."[29]

This was not the first time the extension service and private companies duplicated efforts, nor would it be the last. The extension service felt obliged to offer the germination service because germination tests were an important component of their corn improvement package, and it was difficult to convince farmers of the method's efficacy unless it was demonstrated directly. Like salespeople everywhere, extension agents knew that free samples were persuasive. For the seed companies, testing was also a form of advertising, but for them it served to convince farmers that the company was reliable, honest, and scientific, a "good neighbor" firm that looked out for the local farmer. While there is no evidence that the college and the companies clashed over germination testing, the situation typified the growing ambiguity of roles in the farm community.

The Illinois Crop Improvement Association, 1922–1929

Hackleman's efforts to improve crops in Illinois were not restricted to educating farmers through the extension network.

[29]Hackleman, "Annual Report, 1927," p. 23.

Indeed, his more abiding interests were seed growers and producers. During his time in Missouri, Hackleman had served on the board of the Missouri Corn Growers' Association, which, like the Illinois Corn Growers' Association, concentrated on purifying old strains and maintaining new strains of corn. Such associations, which were organized and directed by seed producers, placed the burden of seed quality, and thus of crop improvement, squarely on the shoulders of seed producers and were partly designed to keep producers honest in providing high-quality seed to farmers.[30]

Seed purity was a recognized problem in agriculture from at least 1900, when the Illinois corn breeders organized. New seed that congressmen distributed to their farming constituents was rarely maintained in its original state, becoming mixed over the years with other seed until the new strain was irretrievably lost. Also, some seed sellers apparently named their ordinary seed after the new strain, making it even more difficult to identify and maintain the new lines. According to Hackleman, a particular strain of wheat distributed by the USDA was later sold under twenty-four different names; it was a practice he considered fairly common.[31]

The difficulty of maintaining pure lines of seed was particularly alarming to experiment station scientists who found, after devoting several years to developing a new strain, that when the seed was distributed to farmers for increase it was frequently "contaminated" or lost after only a few seasons. Because most stations had only a limited amount of farmland available for experimentation, they had little choice but to rely on the cooperation of area farmers who were willing, if not eager, to participate in innovative experiments that seemed to give them an advantage in keeping up with station activities. The solution, then, to keeping seed pure was not changing the distribution mechanism, which, to some, seemed a necessary evil, but rather enforcing the distribution requirements more strictly.

[30]Hackleman, compiler, *History [of the] International Crop Improvement Association, 1919–1961* (1961), pp. 3–6.

[31]Ibid., p. 4.

In response to these concerns the International Crop Improvement Association, a network of state and Canadian organizations that provided field inspection and seed certification to seed and grain growers, was formed in 1919. Its primary function was regulating not only the seed but the conditions under which seed was grown and distributed. Among its tasks were establishing standards for purity and germination, requiring growers to maintain pedigree records, establishing standard tests for identifying strains, inspecting fields where commercial or station seed was grown, and registering and certifying field seed for sale under seal.[32]

In 1921 Hackleman decided the time had come for Illinois to organize a crop improvement association, and in January he called a meeting of prominent farmers and growers to describe the work of the International as well as certification programs that had already begun in other states. Hackleman proposed a general plan of organization in which the Farm Bureau served as the link between seed growers and the college. When the station released a new strain, the Farm Bureau would select a project leader/seed grower who agreed to increase the seed under stringent field conditions. The grower could then sell this seed the next year at a price determined by a committee and could continue as a grower of pure seed as long as the line was maintained. The group approved the plan and appointed an executive committee, comprising three members from the Agronomy Department, one from the farm advisers committee, and two growers. A year later, in January 1922, the Illinois Crop Improvement Association was created.[33]

During its first eight years, the ICIA focused on consolidating membership and refining policies. In 1923 it subsumed the Illinois Corn Growers, and in 1926 the Illinois Corn Breeders agreed to merge as well. With this merger virtually all growers of

[32]Ibid., pp. 14–15.

[33]Alvin C. Lang, *Fifty Years of Service: A History of Seed Certification in Illinois, 1922–1972* (1973), pp. 13–18.

consequence in the state were members of the ICIA. Certification and inspection during these years was intermittent, largely because of the unexpected complexity of the project, especially the difficulty of determining standards of purity for all grains and seeds. Oats, soybeans, and to a lesser extent, clover, were the main crops inspected.[34]

The ICIA was, de facto, the only organization in Illinois that regulated matters pertaining to growing and selling field seed; in effect it took over the college's role as both seed expert and judge. While those Hackleman invited to the first organizational meetings were not, by and large, corn growers, the board of directors soon came to resemble a who's who of prominent Illinois corn breeders and growers. In addition to such respected seedsmen as O. J. Sommer, William Webb, and C. A. Rowe, by the early 1940s representatives of the three big corn companies had sat on the board and on several occasions had held the presidency. At no time, however, did the major seed companies overlap in their tenures; some members felt that these companies wielded too much power as it was.[35]

Membership in the ICIA was all but mandatory for growers who wanted to stay in business. Those who did not participate in the certification program found that farmers discriminated against their seed. For instance, in Tolono, Illinois, the Soybean Club adopted a superior variety of bean in 1923, which farmers in the threshing ring were required to use. Not coincidentally, one of the top seed houses in Illinois, Riegel Seed Company, was located in Tolono, and its head, W. E. Riegel, was president of the ICIA from 1920 to 1932. Growers of certified seed could also undercut

[34]Ibid., pp. 16–18, and see app. F, which tabulates how many acres of crop seed were certified from 1921 on. Records were not available for all years, however, and Hackleman claimed elsewhere that no seeds were certified from 1926 to 1929. See his "Annual Report, 1929," p. 4.

[35]DeKalb was represented in 1923, 1924, 1928, and 1933–36; Funk in 1937–40; Pfister in 1929–32. Lester Pfister's relationship with the ICIA was strained. It appears that he never did certify his seed and in later years was extremely bitter about the way the association was organized.

the price of uncertified seed; in 1929 uncertified beans were selling for $3.50 a bushel, while certified growers sold theirs for $2.50. This maneuver was certain to popularize certified seed as well as to bully unaffiliated growers into joining the association.[36]

For the big companies, who had built up a loyal clientele and did not need certification to succeed, membership in the ICIA was both symbolic and strategic. First, in a community of seed producers as fraternal as those in Illinois, the large companies did not want to seem uncooperative in the pure-seed campaign. Indeed, it was to their advantage to appear magnanimous toward the smaller companies, some of which acted as sellers for them. Second, by participating in the ICIA and in particular by sitting on the board, large companies were able to help design and guide policies to further their own interests.

The role of the college in the ICIA was somewhat ambivalent. In earlier years Davenport had been the guiding spirit behind the formation of various farmers organizations, but things were different by the time Hackleman organized the ICIA. First, Hackleman not only founded the group but served as its secretary-treasurer from 1921 to 1927 and again from 1929 to 1937. Second, the organizations started by Davenport were not nearly as aggressively commercial as the ICIA, which was interested above all in controlling conditions surrounding the sale of field seed. In a sense the ICIA acted as a cartel in which seed dealers regulated themselves with minimal college oversight.

Not everyone was satisfied with the college sponsorship of the ICIA; the most notable detractors were USDA officials. In 1925 George E. Farrell, director of the federal extension program, informed the ICIA that Smith-Lever extension funds could no longer be used to pay the salary and expenses of extension specialists who were "acting as Secretaries of pure seed associations." Farrell's reasoning was that such associations had reached

[36]Hackleman, "Annual Report, 1923," p. 19; "1929," p. 35; Lang, *Fifty Years*, p. 127.

maturity and were commercial rather than educational in character. This directive was certainly aimed at Hackleman, who in his 1925 annual report reassured the agronomy department that the college was "in no way . . . responsible for the activities of the ICIA." But as chief of a college division the entire aim of which was to educate farmers to act with enlightened self-interest, Hackleman certainly was a representative of the college, and his position in the ICIA could easily be interpreted as an endorsement of the association and its goals. It is difficult, however, to understand how Hackleman in the role of ICIA officer was serving in an educational capacity.[37]

Some growers, too, felt that Hackleman's role was inappropriate. One grower of certified wheat complained to Burlison after being excluded from the organizational meetings in 1922. Speaking on behalf of other disgruntled growers, he declared that "the organization as now conducted is contrary to the principles laid down by Dean Davenport and others governing extension work in this state, viz., that farmers themselves shall run their own business and decide their own policies by and with the advice of the University representatives. That the opposite is true in the case of the ICIA is so obvious the most unobservant farmer is able to see it. In this case your Crops [Extension] Division, by hand-picking the officers, dictating inspection fees and all rules, is manipulating a farmer-financed organization contrary to Illinois policies."[38] Despite these criticisms, Hackleman and the ICIA continued their mission during the better part of the 1930s, and the college continued to cooperate nominally. Ultimately the most serious problems the ICIA encountered were defined not by the political issues its very existence created but rather by the complexities introduced with hybrid corn in the late 1930s.

[37]Lang, *Fifty Years*, p. 17; Burlison to Hackleman, 11 April 1937, 8/6/2, box 9, UIA; Hackleman, "Annual Report, 1925," pp. 17–18.

[38]Roscoe Farrar to Burlison, 5 September 1922, 8/6/2, box 9, UIA. The inspection fee was ten dollars for the first twenty acres, plus twenty-five cents an acre above that. Membership was limited to Farm Bureau members, further strengthening the interrelationships between progressive farmers and the university.

The Beginnings of Hybrid Corn, 1930–1935

The period of gradual but sustained growth and diversification in the corn program at Illinois during the 1920s was followed by a decade in which a series of shock waves unsettled the foundations of the college's carefully laid plans. Although corn research appears to have continued steadily, extension and breeding regulation through the ICIA were thrown into confusion by both the economic depression and the commercial introduction of hybrid corn by private breeders. In retrospect, hybrid corn appears to have been more permanently destabilizing than the depression, and in a curious way, the development of hybrids gained strength from the changes in farm and college organization made necessary by the depression.

The first hybrid corn introduced commercially in Illinois was offered in 1929 by Funk Brothers under the name Pure Line Double Cross No. 250. While Funk was jubilant, the college was restrained and made no special note of this innovation. The following year Hackleman began a project with Funk Brothers for testing hybrids and comparing their performance across the state with open-pollinates. At the end of five years, hybrids averaged six bushels per acre more than open-pollinates, an interesting, but hardly overwhelming, demonstration.[39]

Hackleman's prompt study of hybrid performance should not be mistaken for enthusiasm. Indeed, for most of the decade he adhered to his campaign of teaching Illinois farmers how to select good seed corn by field inspection and germination tests. He claimed in 1931 that corn seed that had been selected, sorted, properly stored, and tested on the germinator would yield ten to fifteen bushels an acre more than ordinary seed and called the germination test "the final or acid test to get out all [the disease] that it is humanly possible to remove." Working through the Boys' Corn Clubs, Hackleman compared the yields obtained by the vigilant boys, who adopted virtually all the extension recommendations for corn growing, and their skeptical fathers. In 1931 he

[39]Hackleman, "Annual Report, 1934," p. 26; Hackleman to G. R. Mesenberry, 19 May 1933, 8/6/2, box 9, UIA.

gleefully reported that the boys had averaged ten bushels an acre more than their fathers. Many extension workers considered this a particularly effective demonstration. Also effective was an attack of cephalosporium in 1932, which destroyed 20 percent of the corn crop in Illinois. This Hackleman considered "the best advertisement for culling."[40]

The most persuasive argument for adopting Hackleman's corn improvement strategies, however, was found in the 10-Acre Corn Yield Contest inaugurated in 1930 under the joint direction of the ICIA and the college. For the purpose of the contest the college purchased seed corn from commercial growers, such as Funk, Pfister, and DeKalb, tested the seed, and then sold it at cost to farmers wishing to participate. As time went on they also sold their own hybrids as well as those produced by the USDA and other experiment stations that were suited to Illinois conditions. Farmers would then grow the corn under conditions specified by the college, maintain records on field conditions and harvest, and finally supply the college with these experimental statistics.[41]

The yield contest offered something for everyone. The college, at almost no expense, received experimental information on the adaptability and yield of a wide variety of available corn types, information it had neither the fields nor the staff to acquire alone. The seed companies got free advertising for their corn seed, information about its merits relative to the competition, and a cheap and efficient means of testing new lines that were not yet in commercial production. And the farmers, who saved money buying seed corn from the college rather than a seed house, were able to determine which of the many corn varieties was best suited to their particular conditions.

According to Hackleman, the contest was "a splendid piece of demonstration material, and is calling the attention of the farmer to the fact that corn production is not merely the selection of seed

[40]Hackleman, "Annual Report, 1930," p. 6; "1931," pp. 15, 35, 37; "1932," pp. 11–12.

[41]In 1924 the university and ICIA inaugurated a corn show at which they introduced a new scorecard that popularized "utility," or smooth, corn. The new scorecard and show attempted to turn farmers' attention away from appearance toward field performance.

corn, or merely the planting of corn at a certain time in the moon, but there are many factors which must all be taken into consideration and properly matched if maximum production is to be secured." The contest served as a provocative advertisement for Hackleman's approach to corn growing, a highly visible means of converting the more recalcitrant farmers. One of the test fields in 1933, for example, compared the yield of ordinary open-pollinated corn, a popular certified variety, and a hybrid. The hybrid yielded eighteen to twenty-two bushels an acre more than the ordinary strain, with certified seed corn falling in between the two. This was a rather unfortunate outcome for Hackleman, of course, who was trying to sell "self-help" to farmers. While the test showed that selection could increase the yield four to ten bushels an acre, farmers were probably more intrigued by the even greater yield of hybrids.[42]

Hackleman's efforts to popularize certification also continued to occupy a major portion of his time. Calling the ICIA a "splendid adjunct to the farm crops extension work," virtually "another area of the extension service," he worked to solidify the relationship between the ICIA and the college. Before the ICIA would certify anyone's seed, for example, that seed had to undergo college tests and receive experiment station approval. And as hybrid corn became more popular in the middle to late 1930s, the college, through the ICIA, devised stringent rules to govern the production and commercial release of hybrids. When the ICIA revised its standards for certification in 1936 to include hybrids, it ruled that producers wanting to certify their hybrids could do so only if the lines of which the hybrid was composed had previously been certified. In a four-way cross AB × CD, the resulting hybrid could be certified only if all four inbred lines—A, B, C, and D—had been certified. This rule would create considerable problems for the ICIA and the college later in the decade.[43]

[42]Hackleman, "Annual Report, 1931," p. 41; "1932," pp. 16–20, 30; "How Much Corn Can You Grow?" *Prairie Farmer*, 26 April 1930, 5.

[43]Hackleman, "Annual Report, 1930," pp. 33, 36; Hackleman to O. S. Fisher (OES), 3 November 1932, 8/6/2, box 9, UIA; "Produce Certified Seed," *Prairie Farmer*, 7 February 1931, 16; Lang, *Fifty Years*, pp. 26–27.

Despite Hackleman's enthusiasm, however, the Great Depression was working against him. In 1931 the yield contest drew far fewer entrants than anticipated because the application fee of five dollars was too dear for most farmers. Similarly, few growers were interested in certifying their seed because it too was an extra and apparently avoidable expense. Hackleman admitted that "when corn sells for 10 to 15 cents a bushel it is rather difficult to convince a farmer that he can afford to take time to become more efficient in his corn production." Hackleman, too, suffered from the shrinking economy. The expense of putting gas in the cars and food in the stomachs of project leaders and farm advisers seemed unacceptable by 1932, forcing the extension service to drastically reduce demonstrations and community meetings. By 1935 Hackleman lamented that "during the past several years the production and sale of certified seed has been very discouraging because of emphasis on reduction programs. It has been practically impossible to convince anyone except the most intelligent farmers to spend additional money to secure pure seed of higher yielding sorts."[44]

But the Depression was not Hackleman's only problem. By 1934 farmers had become increasingly interested in the new hybrids, creating a whole new set of problems for the extension service.[45] For more than ten years Hackleman had preached a gospel of self-help: he taught farmers how to identify diseased seed corn in their fields, how to detassel such corn to prevent its perpetuation, and how to build and operate seed germinators. In short, he had worked hard to demystify the scientific components of corn growing, attempting to persuade farmers that significant improvement would result from their own efforts to learn a few simple rules of thumb.[46]

Hybrid corn rendered these hard-won lessons pointless and, in

[44]Hackleman, "Annual Report, 1935," p. 81.

[45]As early as 1930, farm journals were suggesting that hybrids were the wave of the future; see "Hybrid Corn Yields Well," *Prairie Farmer*, 8 March 1930, 14; "Hybrid Corn Stands Up," *Prairie Farmer*, 12 April 1930, 30; "Hybrid Corn Wins Boys' Corn Contest," *Wallaces' Farmer*, 21 February 1931, 249.

[46]Hackleman, "Annual Report, 1931," pp. 9, 41; "1932," pp. 2–3, 10; "1934," pp. 7, 15; "1935," p. 81.

fact, of negative value, for many of the farmers who decided to try the new hybrids persisted in traditional methods of growing corn. Most troublesome was their determination to select corn from hybrid fields just as they had been instructed to do in open-pollinated fields to acquire seed for planting the following season. Corn breeders knew that the hybrids performed well for one season only; second-generation seed, that is, seed obtained from the crop, performed much less well and often worse than the standard open-pollinate. But farmers were not savvy to this fine point, and by 1934 Hackleman had his hands full trying to teach old farmers new tricks: "One of the duties of this division in the future must necessarily be to assist in holding these demonstrations and calling to the attention of farmers . . . that the selection of seed from fields of hybrid corn is a precarious undertaking and in most cases will be practiced with distinct losses."[47]

The second pressing problem with hybrids was that in the early years, the few hybrids on the market were not adapted to the many different growing conditions of Illinois farms. That is, the hybrids had been developed under field conditions that may or may not have been similar to those of the farmer buying the seed. Thus, farmers often discovered in the early years that their own open-pollinated varieties, which had been selected over the years in reference to particular field conditions, performed better and yielded more than the hybrid. Hybrids were so sensitive to differing soils, types of fertility, and climate that no one was able to predict their performance with certainty. Hackleman called adaptability "one of the greatest problems confronting the extension worker in any state of the corn belt now" and advised extreme caution to those choosing to grow hybrids. He told the farm advisers "to caution farmers about stampeding to something merely because it has the name hybrid," but the stampede was unstoppable. In 1934 only .0013 percent of Illinois corn acreage was planted with hybrids; six years later the figure was 77 percent. For Hackleman this transition was anything but smooth.[48]

[47]Hackleman, "Annual Report, 1934," pp. 23, 18.

[48]Hackleman, "Annual Report, 1933," p. 20; "1935," pp. 17–18; "Historical Data for President Kinley," p. 2.

Hackleman's extension problems in the mid-1930s were not unlike those of Davenport at the turn of the century. For both men, the very success of their propaganda and boosterism formed the basis of their later troubles. With varying degrees of success, they persuaded farmers to rely on university authority and goodwill by providing seemingly unlimited service and expertise. Farmers, for their part, formed fairly rigid expectations of the college; in Davenport's time they bombarded the meager station staff with requests for advice and assistance, taking Davenport at his word when he claimed that the college was theirs to command. In Hackleman's day farmers adopted his recommendations for corn improvement so thoroughly that, when faced with hybrids, they were unable to appreciate that the rules of the game had abruptly changed.

Hackleman also had difficulty with the new realities. In 1938, when over three-quarters of Illinois corn acreage was in hybrids, he continued to encourage farmers to grow open-pollinates under the 1930 guidelines. He considered the triumph of hybrids a mixed blessing, but as better-adapted hybrids were developed during the 1930s, he realized that even if yields were ignored, hybrids offered growers several distinct advantages, primarily resistance to disease and insects and stronger root systems. This superiority pulled the rug out from under Hackleman's program, in which successful corn production relied not on the plant itself but on the farmer's informed attention to it. Second, the popularity of hybrids effectively shifted the farmers' interest away from the extension service, which took on a decidedly conservative tone, and toward the seed companies, which had introduced their hybrids under a banner of scientific enlightenment.

The issue here is not who was "first" to introduce hybrids to farmers; the college, after all, was not a commercial firm that wanted to sell a product, and agronomists there did not tend to think in terms of competition with seed companies. Yet on another level, the college—particularly the extension service—and private seed companies were unquestionably rivals for the attention and respect of Illinois farmers. Both wanted to be considered the authority on agricultural matters. Further, there was no clear demarcation between agricultural issues on which each could claim a special expertise. As I suggest in the following chapters, the major seed companies modeled themselves after the college, mimicking its structure as well as its function.

Thus, in this subtle battle for authority between the college and the seed companies, the relatively rapid rate at which farmers adopted hybrids represented a rejection of extension recommendations in favor of the bright promises offered by the seed companies. How the seed companies rose to such prominence and authority and managed to destabilize the seemingly solid research and extension program at the college is the subject of the following chapters.

5

The Uses of Science:

Funk Brothers Seed Company

On a hot summer day in 1920 a group of sixty-five farmers from Arkansas gathered on Funk Brothers Farm outside Bloomington, Illinois. The governor of Arkansas and the former governor of Illinois were there, as was Dave Thompson, secretary of the Illinois Agricultural Association and formerly a farm adviser for the University of Illinois. Like so many other delegations of farmers, scientists, and politicians, they had come to Funk Brothers to observe firsthand the laboratories and cornfields that had come to symbolize an aggressively scientific approach to agriculture. Perry Holden, who, after helping establish Funk Brothers Seed Company in 1901, had become director of the educational program at International Harvester, captured the spirit in his address to the Arkansas group: "Right here you are visiting in my estimation an enterprise that is doing more for profitable farming and for mankind than anywhere else in the United States. No matter whether you are a livestock fellow or a grain man, stock is dependent primarily upon corn. Corn determines our stock, it determines our pigs, and right here we have a movement seventeen years old where they undertake an investigation of our corn. It is remarkable."[1]

And indeed, Funk Farms was fairly remarkable in comparison to other seed companies and even to the University of Illinois

[1]Funk Brothers Seed Company (FBSC), 1920 catalogue, p. 19.

College of Agriculture. Unlike most seed companies, Funk Brothers from the beginning had conducted research informed by the most recent scientific trends and aimed directly at improving corn performance through breeding. A few other companies were also trying to improve corn by selection, but only two or three employed professional agronomists or equipped a laboratory for testing seed. Most companies simply grew and sold high-quality strains of standard varieties. Funk Brothers was also unusual in the acreage it had at its disposal: owing in large part to Isaac Funk's early settlement and land acquisition in Illinois, by 1900 the large brood of Funk brothers and cousins together held over twenty-two thousand acres of prime Illinois farmland. And in terms of social and political presence, the Funk family had long commanded attention. From the 1820s on, Funks were rather persistently in the public eye as bank presidents (three), bank directors (two), members of the state legislature (four), congressmen (two), University of Illinois trustees (two), and a mayor. In the agricultural arena as well Eugene Funk, founder of the company, was prominent. He was a member of the Illinois Corn Growers' Association, a founding member of the Illinois Corn Breeders' Association, a life member of the American Breeders' Association, president of the National Corn Association during its active years, 1907–1914, and a member of the University of Illinois Farm Crops Advisory Board from 1920 to 1940.[2]

By 1920 Funk Brothers also compared favorably with the college in terms of its research program and public presence. Eugene Funk had been interested in corn breeding nearly as long as college agronomists. In 1893 he had developed a new strain of corn. He had also cooperated vigorously with Hopkins's oil and

[2]For information on the Funk family, see Helen Cavenagh, *Funk of Funk's Grove: Farmer, Legislator, and Cattle King of the Old Northwest, 1797–1865* (1952), which is primarily about Isaac Funk, the family patriarch; and Cavenagh, *Seed, Soil, and Science*, which is primarily about E. D. but also has a good deal about the entire second and third generation of Funks. A brief description of Funk achievements can be found in FBSC, 1925 catalogue, pp. 1–4.

protein selection experiment, tested the new Mendelian theories on corn, and redefined standards of corn quality following a series of experiments on the relation between appearance and yield. After the company hired Purdue agronomist J. R. Holbert in 1917 and acquired USDA sponsorship for his extensive studies of corn disease, the emphasis on scientific agriculture at Funk Brothers was even more pronounced. In seed catalogues, public addresses, and interviews, the company's commitment to the systematic, scientific investigation of corn production—in contrast, incidentally, to its competitors' seeming lack of such an approach—was a recurring and compelling theme.

Curiously, at a time when Funk Brothers was using science to bolster its image, the college was using practical issues to enhance its own. Both were interested in capturing the farmers' attention by trying to balance their traditional area of expertise with the expertise of the other. For the college, which was constantly on guard against losing the farmers' support, a scientific image was both an asset and a liability. The image attracted the more educated and affluent farmers, who looked to the college for advice and assistance, but to many others the college seemed out of touch. These farmers were suspicious of "book farming," academically based agriculture that relied too much on hothouse science and too little on practical experience. With the corn program the college played down the more complex and esoteric elements of genetics and physiology when addressing and educating farmers. Agronomists sensed that farmers were only too inclined to view college advice as overly scientific and thus were anxious to demonstrate that the college could also be practical and, quite literally, down to earth.

For Funk Brothers the problem was different. Seed houses were expected to be reliable rather than innovative, and Funk Brothers decided at the outset that infusing the business with the trappings and benefits of scientific agriculture would set it distinctly apart from more traditional seed companies. Funk Brothers developed and promoted a scientific image that, in some cases, made the company appear more scientific than the college. Many of its research and promotional efforts were obviously patterned after land-grant-college strategies, and the attention Funk Brothers commanded from prominent scientists was truly impressive. Ultimately, it appears, Funk Brothers was more successful in attempting to resemble an experiment station than the station was in pretending to be a farm.

Still, Funk Brothers was not strictly a research laboratory. It was a seed company, and it would be misleading to suggest that its intense interest in corn grew from fundamentally scientific rather than economic concerns. Company research contributed a great deal to the broader questions of inheritance, but those contributions should be considered the icing, not the cake. All the experiments conducted by the company and, later, the federal field station, were guided by the question "how can the yield and quality of corn be improved in a profitable manner?" That the answer to this question involved a more theoretical component was a happy circumstance for all concerned.

The Business of Corn

Eugene Duncan Funk was born into the emerging Funk dynasty in 1867. His father, Lafayette, farmed on the family's land as did his uncles. When Eugene was eighteen he entered Andover Academy and three years later went on to the Sheffield Scientific School at Yale, where he studied the natural sciences. In 1890, instead of graduating from Yale, he decided to visit Europe with his cousin I. Linc Funk. During their four months in Europe, the two young men saw the usual sights but paid particular attention to agricultural conditions. The highlight of their trip, reportedly, was a tour of the famous Vilmorin estate in France. At about this time the Vilmorins were receiving international recognition for their agricultural enterprise because they had increased the sugar content of beets through selective breeding. Eugene later remarked that the visit to the Vilmorin estate was the inspiration behind his own seed business.[3]

After his European tour, Eugene returned to Illinois full of enthusiasm for scientific agriculture. Like many others interested in corn breeding, he read W. J. Beal's 1886 report on crossing corn varieties to increase yield, and in 1892 he attempted his own experiment. Spring planting that year was delayed because of an unusually wet spring, so Funk planted an early-maturing variety called Early Murdoch, obtained from the University of Illinois. The following year he got another early variety—Pride of the North—which had been developed at the Minnesota experiment station by C. P. Bull and Willet M. Hays. Funk crossed these two strains and developed Funk's 90-Day, a varietal cross that the company grew and sold until the late 1930s.[4]

Encouraged by the development of Funk's 90-Day, Eugene and fourteen other Funks who had all been farming separately decided in late 1901 to combine their land and resources for the purpose of growing and selling corn and other field seeds. As president of Funk Brothers Seed Company, Eugene strove to anchor the company at the beginning upon firm scientific ground. In searching for someone to help set up the research program, he turned to breeders at the University of Illinois. When A. D. Shamel declined Funk's offer, Funk hired Perry Holden, who was finishing his first year at a beet-refining plant.[5]

[3]A description of Eugene's diary entries during his college days and European tour can be found in Cavenagh, *Seed, Soil, and Science*, pp. 55–67. While this book contains a great deal of information on Funk, it is nearly devoid of critical analysis. As the only published work on Funk, however, it has served, provisionally, as an underpinning for this chapter.

[4]FBSC, "A History of Hybrid Corn," 1940, p. 4; Cavenagh, *Seed, Soil, and Science*, p. 85; Colin Kennedy, "Funk Name Long Associated with Seed Corn Improvement; Extensive Research Conducted," *Chicago Daily Drovers' Journal*, 30 September 1940, 6. Holbert later wrote that Funk's 90-Day was not widely used because the ears were too small to produce a good yield. It was reliable, however, when wet spring weather necessitated late planting; see Holbert, "Funk's 176-A Story" (manuscript, 1945, Funk Files), pp. 3–5.

[5]FBSC, "Supplemental Data Pertaining to Application for Relief under Section 722 of the Internal Revenue Code for the Fiscal Years Ended June 30, 1941, 1942,

During its first ten years, corn research at Funk Brothers reflected the scientific trends prevailing at the Illinois experiment station and signified Funk's attempt to apply the new scientific techniques to practical problems in corn production. Funk was very interested, for example, in Hopkins's oil and protein selection experiments, and in 1902 he began selecting corn that had a chemical composition suitable for feeding livestock. Although the company was not yet equipped to conduct the chemical analyses itself, Funk struck a deal with officials at Illinois Wesleyan College in Bloomington. Funk approached R. O. Graham, a Johns Hopkins–trained chemist who was professor of chemistry at Wesleyan, and proposed that Funk Brothers would endow a new chemical laboratory at the college if Graham would analyze Funk's corn strains. Graham and the college agreed, and Graham hired H. H. Love to assist him with his work.[6]

For three years the Wesleyan laboratory provided Funk with chemical analyses, but in 1905 Funk decided to abandon the project. Although the analyses and selection experiments had allowed Funk to alter the oil and protein content of corn significantly, as Hopkins himself had done, "the buying public did not recognize the added commercial value." Further it appears that the University of Illinois was somewhat distressed by Funk's relationship with Wesleyan; in 1905 the university offered to provide Funk with chemical analyses of his corn at no charge and also created a position at the station for H. H. Love.[7]

Although the aim of selling specialized corn seed had proven futile, the production method it introduced was not. When the oil and protein work was abandoned, the ear-to-row selection method was retained because it provided a way to identify both male and female parents, an advantage lacking in traditional

---

1943," pp. 1–8. Section 722 describes rules for companies wanting to average their income, a strategy first available in 1941; Frank Gillespie to Funk, 9 June 1934, Funk Files; Cavenagh, *Seed, Soil, and Science*, pp. 88–89.

[6]Cavenagh, *Seed, Soil, and Science*, pp. 92–93, 98; Leon Steele, "The Founding of Funk Seeds," 1983, [p. 4] Funk Files; Crabb, *Hybrid Corn Makers*, pp. 25–26.

[7]Cavenagh, *Seed, Soil, and Science*, p. 108, quoting Eugene Funk.

mass selection or even in the more controlled varietal crossing techniques. From a practical point of view, ear-to-row breeding was decidedly superior because it provided the breeder with a pedigree for corn that was useful both in repeating particularly advantageous crosses and in advertising the "scientific breeding" behind each strain.[8]

Funk Brothers began ear-to-row breeding in 1901 under the direction of J. Dwight Funk, Eugene's cousin. J. D. was keenly interested in the science of breeding and attentive to both theoretical and practical issues of corn inheritance. His experiments were clearly influenced by the new Mendelian ideas, which, though he understood them only imperfectly, provided a loose array of guiding principles. After two years of ear-to-row tests he reported to the company, "It is an established fact that there are certain laws of nature governing the propagation of characteristics in plants. Certain individuals reproduce certain characters throughout their line of descent while others yield in their descendents [*sic*] to the influence of the fertilizing plant and result in innumerable different types. In the selection of the so-called dominant types lies great development of the different varieties of corn."[9]

In 1903 J. D. planted thirty "breeding blocks" of three to five acres each according to the ear-to-row method. After selecting his best eighty to one hundred ears of corn, he shelled and planted half the kernels on each ear, assigning a number to each row as well as to each remaining half ear. These numbers were entered in a record book, as was a description of the peculiarities of each ear. J. D. considered ear appearance important not because he wanted to breed ears that were considered attractive but because he wanted to obtain uniformity of kernel type. By selecting ears that most resembled the mother ear, he obtained more uniform ker-

[8]FBSC, "Ten Years of Corn Breeding," 1913–1914 catalogue, pp. 2–9.

[9]Agronomist's Report, May 1903, Funk Files, p. 2. J. D. and Eugene joined the American Breeders' Association in 1903, and J. D. contributed several papers to its meetings.

nels, which tended to plant and grow more easily than kernels of different shapes and sizes. As pollination time neared, he also detasseled corn that exhibited negative characteristics, often destroying entire rows if the characteristic was prevalent.[10]

The following autumn J. D. determined that out of 2,100 ear rows, only 121 were worthy of the designation "Champion Rows"; from these he selected both breeding ears for the next year and superior ears for commercial increase and sale. J. D. was very optimistic about the potential of this method for gradually improving all the company's corn: "The plan is to make such selection that the strains of the different varieties that give the greatest yields will finally be planted on our entire acreage, keeping the blood lines pure and thus establishing the only lines of pedigreed corn for sale in the world. It is my plan to so arrange and select the seed for next year and that the catalogs [*sic*] for another year contain a clear pedigree of these high-yielding lines of corn similar to a catalogued pedigree of pure line livestock." But J. D. was also aware that not every promising ear would turn out to be superior. Indeed, by 1903, five thousand ears had been tested, but only two gave a consistently high yield, the rest performing well only sporadically. He hypothesized that in these inconsistent strains the yield "character" was insufficiently fixed or purified, whereas in the two high-yielding strains the character was constant. J. D. therefore questioned H. J. Webber about the genetics involved, that is, whether "these two strains are the result of mutations."[11]

[10]Cavenagh, *Seed, Soil, and Science*, p. 90; FBSC, "A History of Hybrid Corn," p. 6; E. D. Funk, "Commercial Corn Breeding" (address presented to the Congress of Experiment Stations and Colleges of Agriculture, Louisiana Purchase Exhibition, 8 October 1904, Funk Files); [J. Dwight Funk], Agronomist's Report, 14 November 1903, Funk Files; J. Dwight Funk, "Commercial Corn Breeding," *Proc. ABA*, 1903, 29–33.

[11]J. Dwight Funk, "Commercial Corn Breeding," p. 31; Agronomist's Report, April and November 1903, Funk Files; Cavenagh, *Seed, Soil, and Science*, pp. 97–98; Kennedy, "Funk Name," p. 6.

In these early years Funk Brothers also began to detassel plants to control pollination and prevent inbreeding. J. D. noted that at the experiment station Hopkins had studied the effect of detasseling on yield, and the results showed that detasseled stalks yielded from one and a half to six bushels per acre more than tasseled corn. The procedure was simple enough: rows of strains A and B would be alternated in the plot; then all the A rows would be detasseled. At harvest, then, the breeder would know that all the A rows had been pollinated by the B strain, and all the B rows had been self-pollinated. While this procedure was considered detrimental to the B rows, which were usually discarded, it provided a pedigree for the A rows. An even more definite means of controlling the parentage relied on hand-pollination. Although crossing strains by the ear-to-row method did increase the yield, after a few years no further improvement could be attained. Funk attributed this ceiling on improvement to the imprecise lineage of the pollen parent. Although the pollen strain was known, the pollen ear was not. J. D. maintained that although all pollen parents showing negative characteristics at pollination time were routinely detasseled, some of the remaining pollen parents would in fact not perform well. The only way to identify which plants were contributing these negative qualities was to hand-pollinate all crosses and enter each cross in the record book. According to J. D., "any plant breeder who has really accomplished results of utility has increasingly controlled the influence that is exerted by both parents upon the progeny."[12]

The ear-to-row work also brought to light another problem that had been worrying Eugene for some years: the relationship between ear appearance and yield. In 1890 Funk and fourteen other seed producers began to gather informally to discuss ways of improving corn yield. This group, which came to be known as the

[12]J. D. Funk, "Practical Corn Breeding on a Large Scale," *Proc. ABA*, 1906, 89–93; Agronomist's Report, May and June, 1904, Funk Files; Cavenagh, *Seed, Soil, and Science*, pp. 100–113.

Illinois Corn Breeders' Association, decided to devise a scorecard that would serve as a guideline for producing uniform types of high-yielding corn. One of the main features of this scorecard corn was its emphasis on roughly indented kernels, a characteristic that most farmers and growers favored. Funk began concentrating on producing rough corn that conformed to the card, but became suspicious of the criterion when, after four or five years, the yields of rough corn "went all to pieces." He decided to conduct an experiment to test the relative merits of smooth and rough corn. Using the ear-to-row method on seven types of standard Leaming corn, for seven years he selected corn that ranged in kernel type from very smooth to very rough. In nearly every case the smooth corn outyielded the rough.[13]

While from a scientific point of view Funk's discovery was exciting, from a commercial standpoint it was troubling. Not only had the Corn Breeders' Association agreed to produce only rough corn, but it had worked to persuade farmers to buy and select for rough corn. Indeed, the scorecard was used at corn shows to reward the efforts of those farmers growing corn that most nearly conformed to the rough ideal. In 1906 Funk wrote to his fellow breeders suggesting that the scorecard was out of date and out of line with the scientific facts. He pointed out that dairymen, who also had a scorecard ideal, nevertheless paid more attention to weight and the results of the Babcock test when selecting breeding cattle. Finding little correlation between ear appearance and yield, Funk suggested the formulation of a new scorecard that considered yield as a prominent factor. But he admitted that redefining the ideal would not be easy: "If the public is to be radically educated along other lines and insists on demanding scorecard corn, then there is little or no inducement for the corn breeder to keep up a system of expensive and systematic records, chemical analyses, or anything but the selecting of show ears."[14]

[13]Cavenagh, *Seed, Soil, and Science*, pp. 216–217; E. D. Funk, "The Search for Better Corn" (manuscript for an address, not delivered, in St. Charles, Illinois, 28 October 1938, Funk Files), pp. 5–6; FBSC, "History of Hybrid Corn," [p. 5].

[14]E. D. Funk to corn breeders, 7 June 1906, Funk Files.

Other breeders agreed. G. O. Sutton remarked that his fields produced excellent yields although the ears were too large for the scorecard ideal. Hopkins considered yield before the scorecard as well, his tone suggesting that no one took the scorecard seriously anyway. According to Eugene Davenport, "All these scorecards are the result of opinion and taste rather than of exact studies to determine what type is most prolific or useful." Albert Hume, however, pointed out the difficulties of including yield as a factor in corn shows where the scorecard reigned. First, average yield was highly inconsistent from one acre to another and from one farm to another; therefore it was difficult to measure the yields of one farmer against those of another. Second, no single ear could be considered representative of a row or an acre; indeed, while yields referred to populations, scorecards referred to individuals. Given this basic incompatibility, it is not surprising that neither the Corn Breeders' Association nor any other organization tackled this problem immediately.[15]

Despite these difficulties, Funk Brothers effectively abandoned the scorecard ideal and began breeding for yield rather than appearance. In the 1913 catalogue Funk devoted considerable space to an explanation of the corn-breeding methods practiced at Funk Brothers, including a restrained criticism of the scorecard ideal. Suggesting that "the commercial corn breeder should not allow himself to become too greatly absorbed in ideal ears, unless these can show for themselves by authenticated records prepotent powers of reproduction," he further noted that "personal experience has abundantly proven some of our highest yielding strains of corn today are anything but ideal ears from the standpoint of the scorecard." Eventually all of Funk's commercial strains were bred to conform to this smooth ideal. Funk's 90-Day, Yellow Dent, and Leaming were all promoted as "Utility" corn, a name meant to indicate that smooth corn was more productive than scorecard

[15]G. O. Sutton to Funk, 7 September 1906, C. G. Hopkins to Funk, 12 June 1906, E. Davenport to Funk, 15 June 1906, A. D. Shamel to Funk, 19 June 1906, and Albert Hume to Funk, n.d. [concurrent with others], Funk Files.

corn. By 1921 smooth corn officially replaced rough corn on the Illinois scorecard.[16]

Commercial Selection and Breeding

By 1914 the breeding program at Funk Brothers was well under way, as was its advertising program. Year after year, the annual spring catalogue was full of details describing the fields; the breeding blocks; the standards used for selecting "champion strains"; methods of breeding, germinating, and increasing seed; and most of all, a sense that because of its scientific ingenuity, Funk Brothers Seed Company could promise healthier, higher-yielding strains of corn. The company printed scores of testimonials from farmers that handily verified the company's claims, and many farmers reported regret that they had not used Funk all along for their field seed. A hypothetical conversation published in the 1921 catalogue typified the Funk Brothers selling style:

> Corn Grower: What does it cost an acre to plant Funk's High Yielding Seed corn?
> Answer: It will average between 65 and 75 cents.
> Corn Grower: Can you not afford to sell this seed for less money?
> Answer: Not unless we cut down our extensive corn breeding work to increase yields and spend less on the selection and care of our seed.
> Corn Grower: We can save 15–20 cents per acre by planting our own corn or buying cheaper seed.
> Answer: Yes but the chances are Funk's well bred seed will yield 10 or 15 bushels more per acre. . . . Thousands of other farmers have had similar results.
> Corn Grower: These men seem to back up what they say and I want enough seed to plant 100 acres. Let me see: 12 and

[16]FBSC, 1913–1914 catalogue, p. 7; Cavenagh, *Seed, Soil, and Science*, p. 407.

> 1/2 bushels at $5 is only $62.50. I am willing to pay 15 or 20 cents more an acre for extra good seed because if I get 10 bushels or more to the acre it means a net profit of over $6 an acre or more than 500% on my investment.[17]

By the mid-1910s, Funk had earned a reputation not only as a progressive seed house but also as a research site for innovative agricultural practice. In 1914 George Shull wrote to Eugene Funk from Berlin with suggestions for a research project involving "direct hybridization" rather than selection as a method of improvement. "If you find by careful experimentation that some form of direct hybridization will give higher yields with better quality and so forth, than your present methods, your statement to this effect would be accepted by your constituents quite as readily as they now accept your statement regarding the value of intelligent selection on the basis of performance records in the ancestry." Funk admitted that he was hesitant about changing the direction of his research, but Shull pointed out that "it is not the method which you use but the name of Funk Brothers that has become impressive to the intelligent farmer." And indeed, Shull suggested that Funk should pursue this line not only for his own interest but for the broader interest of scientists not as well endowed: "You could afford at least to carry out some fairly extensive experiments to test the value of hybridization methods, and I believe that you owe it to yourselves and your constituents to undertake experiments along this line, to be guided as your work has always been, by the purely practical consideration of securing the largest possible yields."[18]

While it is not entirely clear what Shull meant by "direct hybridization," it seems likely that he was referring to varietal crossing by means of individual hand-pollination. The following year Funk did experiment with a three-way varietal cross and, rightly

[17]FBSC, 1921 catalogue, p. 22.

[18]Shull to Funk, 17 January 1914, Funk Files; Funk, "The Search for Better Corn," quoting Shull on p. 8.

or wrongly, seemed to feel that this was what Shull had in mind. Funk's Tribrid, first sold in 1911, was an attempt to combine the feeding value of Leaming with the yield of Funk's Yellow Dent, the company's most popular selected strain, and the early maturity of Funk's 90-Day. After a few years, however, Funk quit producing Tribrid because its performance and popularity seemed insufficient to warrant the cost of producing it. Although in later years it was rather inaccurately labeled the first hybrid sold by Funk Brothers, Tribrid was in essence a varietal rather than an inbred cross.[19]

Nonetheless, like Funk's 90-Day before it, Tribrid reflected Funk's early interest in identifying corn by specific characteristics that, when combined, could result in the economy of purpose hybrids came to symbolize. If strains consistently expressing any of the many characteristics valued by farmers—early maturity, strong roots, high yield—could be identified, then in theory such strains could be isolated for further selection and increase. Because no single varietal or, later, inbred strain carried all or even several of these favorable characteristics simultaneously, commercial breeders concentrated most of their efforts on somehow combining the characteristics of different strains into one strain by means of breeding. In both Funk's 90-Day and Tribrid, Funk manipulated existing characteristics and recombined them more or less successfully, establishing his reputation as a knowledgeable breeder. And his single-minded focus on improved corn performance allowed him to achieve results of practical benefit much sooner than breeders at the Illinois station.

Of even greater importance for Funk's breeding program, however, was the appointment of J. R. Holbert as agronomist at Funk Brothers in the fall of 1915. Together Holbert and Funk represented an ideal combination of qualities for a young seed company eager to make its mark. Whereas Funk had received some early scientific training at Yale and had kept up with most of the

[19]Very little has been written about Tribrid; see Cavenagh, *Seed, Soil, and Science*, p. 228; FBSC, 1917 catalogue, pp. 10–11.

scientific literature, his real strength lay in his practical experience with field practice and crop improvement. Holbert, who had just received his B.S. in plant pathology at Purdue, had a strong academic background and training in crops and especially crop pathology. At his own instigation, Holbert had spent the summers of 1913 and 1914 working on Funk Brothers farms and in the summer following his graduation had gone to Minnesota as an agent of the USDA to participate in a cereal disease survey. During the summer Holbert weighed his professional options. Not only had Funk offered him the job, but an "eminent plant scientist" (probably G. N. Hoffer) had a spot for him at Purdue, the USDA had a permanent position for him, and several experiment stations had offered him research fellowships.

Holbert's decision to reject the more traditional positions and join Funk Brothers indicates the peculiar opportunities available to plant breeders during this formative period of plant breeding research. He made his decision on the advice of both John Parker, his supervisor on the Minnesota project, and H. K. Hayes, who, after studying with E. M. East at Connecticut, had just arrived in Minnesota to begin a breeding program at the experiment station. Hayes, of course, was one of the few academic experts on experimental corn breeding at this time, and he urged Holbert to use Funk Brothers as a laboratory for inbreeding and crossing corn.[20]

Holbert's first assignment at Funk Brothers, however, was concerned not with inbreeding but with selection. In what turned out to be Funk's last major selection experiment, Eugene Funk instructed Holbert to select field seed by following a stringent and, to Holbert's way of thinking, perplexing standard. Using Funk's Yellow Dent as a basis, he was to select only good strong ears on standing stalks, on which the ears were at a convenient height, and that were surrounded by other good plants. Funk did not want selection based on any scorecard standard that considered

[20]"James R. Holbert," *American Men of Science*, 9th ed., p. 524. Crabb, pp. 103–108; Holbert, "Funk's 176-A Story" (manuscript, 1945, Funk Files), pp. 1, 3; Hayes, *Professor's Story of Hybrid Corn*, p. 31.

kernel type or ear length, factors that Holbert, as an official corn show judge in Indiana, would certainly have included. Following Funk's standards, Holbert found that only a few hundred ears of corn met the requirements, enough to plant forty acres the following spring. Virtually all the ears were of the smooth-kernel type, and all tested well on the germinator.[21]

After selecting for such corn for two more years, Holbert ended up with the aforementioned Utility corn, which Funk Brothers began selling under existing corn names in 1917. Funk and H. H. Miller, general manager of the company, were eager to endow the new strain of Yellow Dent with an image that would signify Funk's commitment to scientific approaches to crop improvement. After rejecting the simple name Funk's Utility, Miller suggested that the corn be identified by a number "to signify the experimental work back of it." Holbert obliged by providing the coded pedigree number for the strain—176-A—a name that Miller especially liked because "that A has selling value."[22]

In 1916 Holbert also began inbreeding some of the promising strains that would later make up 176-A. In 1918 he was ready to start crossing the inbreds, as Hayes had suggested, but severe weather destroyed most of his inbred stock. Finally in 1921 Holbert produced a significant number of inbred and single crosses in the field, and the following year Funk Brothers began selling small quantities of Pure Line Double Cross No. 250. It was not advertised in the catalogue, however, until 1928.[23]

[21]Holbert, "Funk's 176-A Story," pp. 2–12; FBSC, "History of Hybrid Corn," p. 11; Crabb, pp. 110–117.

[22]Although Funk Brothers may have sold small quantities of 176-A in 1917, it was not sold through the catalogue until 1921; this common practice with new lines makes it difficult to date "introductions." In 1921, 176-A was also called Yellow Dent; in 1924, it was called Utility Type Yellow Dent 176-A.

[23]Holbert, "176-A Story," pp. 12–13. Both FBSC, "History of Hybrid Corn," and Cavenagh claim that 250 was featured in the 1926 catalogue, but I have failed to locate such a reference. Rather, I find it first mentioned in the 1928 catalogue, pp. 2–3; see FBSC, "History of Hybrid Corn," pp. 12–13; FBSC, "Historical Background to Research Acres," 1950, Funk Files, p. 3; Cavenagh, *Seed, Soil, and Science*, pp. 228–229; Crabb, pp. 117–118.

The inbreds Holbert developed had scientific as well as commercial importance. Scientists were already aware that inbreeding served to concentrate plant characters by reducing them to their most elemental forms, but such an experiment had never been conducted on a large scale. When plant breeders such as Hayes urged Holbert to join Funk Brothers, it was this scale of experimentation that seemed most alluring and represented a significant advantage in breeding work. The sheer number of inbreds in the field provided a dramatic verification of what breeders suspected, and in 1921 especially, experts from around the country flocked to Funk Brothers to see for themselves. Among those visiting were Hayes, Donald Jones, commercial breeders Henry A. Wallace, Lester Pfister, and C. L. Gunn, and university agronomists T. A. Kiessellbach (Nebraska) and L. E. Call (Missouri). What they saw was that while many inbreds were uniformly vigorous, most were uniformly worthless. And the crosses between them were equally striking. "Some crosses were found to have kept only a part of them. Some were found to have developed new weaknesses. Some were found to have combined certain of their vital forces into new and surprising quality and yield."[24]

The fields of inbreds and crosses also suggested that the double cross might be the key to commercial production of hybrid corn. While in 1918 and 1920 Donald Jones theorized about this possibility in academic publications, Funk demonstrated it on his farm. Indeed, Funk provided Jones with the large-scale experimental evidence he had lacked. Jones was excited by this affirmation of his ideas and in 1921 wrote to Funk about their mutual interests. He included a preprint of his 1921 paper and pointed out, rather gratuitously, that inbreeding and crossing were of little use unless practiced on a large scale because such a low percentage of plants actually demonstrated a healthy concentration of the desirable characteristics. Since "the results depend upon the number of plants worked with," his aim was to "interest [Funk]

[24]Crabb, pp. 121–133.

... in this method as a purely business proposition." Jones enthusiastically proposed that inbreds could be sold directly to farmers to cross-pollinate themselves (an idea that station breeders would also briefly consider in the 1930s). While Jones was not interested in leaving the Connecticut station, he felt that "it would be possible to make arrangements with the authorities of this station whereby I could lay out a plan and give enough personal direction to see that it was properly put through provided there was sufficient inducement."[25]

But by 1921 Eugene Funk did not require the "personal direction" of an eminent plant breeder. He was already producing double-cross seed with Holbert and had nothing to gain by aligning himself with Jones. Together Holbert and Funk commanded the attention of most serious commercial and university corn breeders. They watched Funk's work with increasing interest partly because they were intrigued by the hybrid research itself and partly because in 1918 Funk had acquired the sponsorship of the agricultural institution they most respected—the USDA.

The USDA and Funk Brothers, 1917–1935

The USDA and Funk Brothers had cooperated intermittently since early in the century. For example, in 1904 Jesse Norton, an oat expert for the Bureau of Plant Industry, established several breeding plots at Funk Brothers in an effort to improve the oat plant through breeding. Likewise, Charles Brand from the USDA experimented on red clover at Funk Brothers in 1905–1906. In 1907–1908, while studying moisture content in corn at the company farms, Laurel Duval perfected his moisture-testing machine. Funk's name became even more familiar in Washington following his appointment by President Woodrow Wilson in 1917 to the Committee of Twelve, whose task it was to set the

[25]Donald F. Jones to E. D. Funk, 9 February 1921, Funk Files.

wheat price for the 1917 crop. Thus, it is fair to say that by 1917 the USDA and the company had established a mutually beneficial relationship that allowed them to pool their resources. When Eugene Funk suggested that a federal field station be established on his farm, the USDA was predisposed to consider the proposal in a favorable light.[26]

The precise details of the origin of this cooperative arrangement remain unclear, but Funk and BPI officials worked out their agreement in late 1917 and 1918, and the station began receiving federal funding the following July. The BPI "borrowed" Holbert from Funk Brothers, undertaking to pay his salary and obliging him to conduct himself as a federal employee rather than a private breeder. In most respects, however, Holbert's relationship with the company remained the same: though technically a federal employee, Holbert conducted all his research at Funk Brothers, and the company did not replace him when the federal arrangement began. Indeed, it was nearly impossible to distinguish between Holbert's company research and his federal research throughout the twenty-year partnership.[27]

Funk Brothers' arrangement with the USDA was further enhanced by academic support. One might have expected the USDA to designate the Illinois station as the academic arm of the project because of its physical proximity to Funk Brothers and its long cooperation with the company. Nevertheless, the Illinois station was not included. Officials at the USDA felt that the station staff was not strong enough in pathological investigations, and in fact, in 1917 the station was only starting to build a comprehensive

[26]Frank Gillespie (congressman) to Funk, 9 June 1934, pp. 4–6; Cavenagh, *Seed, Soil, and Science*, pp. 120, 133–134.

[27]Secondary information on this cooperative arrangement is exceedingly sparse; it is merely noted in Crabb (p. 120) and Cavenagh, *Seed, Soil, and Science* (p. 384), although the historical significance of the arrangement surely deserves a more coherent treatment. A. C. True noted that in 1913 there were eighteen federal field stations in nine states, which together carried on 60 percent of the federal agricultural research, often in cooperation with the state experiment stations. See A. C. True, *A History of Agricultural Experimentation and Research in the United States, 1607–1925* (USDA Misc. Publ. 251, 1937), pp. 198, 225.

program in corn research and development. Purdue's experiment station in Lafayette, Indiana, was chosen instead, largely because of its strength in plant pathology but also because it was Holbert's alma mater. Holbert's mentor at Indiana, G. N. Hoffer, was put in charge of the laboratory and greenhouse experiments, and Holbert frequently visited him to compare and correlate research findings.[28]

Although cooperating with private companies was not uncommon for the USDA, the policy was problematic. Certainly, the government was concerned about arousing public suspicion of collusion or favoritism. The delicacy of the arrangement was apparent as officials at the Office of Cereal Investigations negotiated with Funk Brothers in late 1917. Referring to a recent USDA ruling that forbade "bureaus and offices to cooperate with any corporation," OCI pathologist H. D. Humphrey suggested that the OCI would be unable to cooperate with Funk Brothers except by employing Holbert "to look after certain experimental plots we shall have at Bloomington." Eugene Funk, incensed by Humphrey's bureaucratic naïveté, responded: "Now either this is a new arbitrary and impractical ruling of the Department, or there is something wrong somewhere, for our seed company has been cooperating with various branches of the Department for the past sixteen years." Funk's objection was forwarded to K. F. Kellerman, associate bureau chief, who explained that the ruling was actually designed to prevent the application of Rockefeller Foundation funds to the Bureau of Education. Although annoying and generated by circumstances irrelevant to the USDA, the ruling did indeed make "direct connection of the work with stations under any kind of cooperative arrangement with any company or corporation . . . impossible." But he continued, "This does not mean . . . that we are not anxious to continue the informal cooperation that is not complicated by financial arrangements." And thus a deal was struck whereby the BPI employed Holbert to

[28]Cavenagh, *Seed, Soil, and Science*, p. 386; C. R. Ball to W. A. Taylor, 24 February 1922, 54/2/50, NA.

"look after" its experimental fields in Bloomington on Funk Brothers Farms for the next twenty years.[29]

Another difficulty arose from the conflict between the very different goals of public and private researchers. Funk's primary objective was to obtain practical information on corn improvement—knowledge he could use in business—on a much larger scale and at a faster pace than he could manage alone. Both the USDA and Purdue, however, were far less interested in immediate practical findings than in long-term contributions to agricultural knowledge. The tension between these two expectations—commercial and scientific—surfaced as early as 1917. The first incident occurred in late September after Purdue officials had taken photographs of the corn-breeding blocks at Funk Brothers. As the company prepared its seed catalogue for the next season, Funk decided he would like to include some of these photographs, and wrote to C. G. Woodbury at Purdue requesting copies. Woodbury, who found the request "embarrassing," objected that "the pictures were taken for scientific purposes" and declared that "it would be unfortunate from the standpoint of the government work in progress with the company to have these prints published at this time." Perhaps Woodbury was reluctant because the USDA had not yet acquired congressional authorization to participate in the project. Yet all participants seemed to consider such authorization a mere formality, as evidenced by their discussion of "the cooperative work just being started" nine months before the authorization.[30]

Funk's response to Woodbury is also instructive. Pointing out that "it would not be to our advantage as seedsmen" to show detailed photographs of the breeding blocks, Funk wanted only those photos showing a general view of the fields. "The average farmer would not be able to comprehend or grasp the meaning of

[29]H. D. Humphrey to Funk, 19 November 1917, Funk Files; Funk to Humphrey, 8 December 1917, Records of the BPI, Office of Cereal Investigations, 54/31/61, NA; H. K. Kellerman to Funk, 10 December 1917, Funk Files.

[30]C. G. Woodbury to O. P. Tieman, 28 September 1917, Funk to Woodbury, 3 October 1917, Funk Files.

our work if the book was filled with pictures along scientific lines, and our competitors (if there are any (?)) would reap the benefit of our knowledge at our expense." Anticipating the release of new lines to the market, Funk wanted to provide some evidence, no matter how scanty, of the scientific program that would soon bear the fruits of research. Funk hoped to whet the appetites of his clientele for improved seed corn so that there would be a ready market when it was introduced.[31]

The business-science tension reappeared a few months later when Funk requested permission to report some of the new findings to farmers. Kellerman was unenthusiastic. "It seems to me," he explained, "that to merely call attention to the difficulties of corn planting is undesirable . . . especially since it is impossible for us to give very satisfactory advise [*sic*] as to what the farmer should do." Kellerman, like most crop scientists, considered the results of experimental field trials provisional at best. He assumed that long-term verification trials would be necessary before such information could be considered at all useful and thus reportable to farmers. Funk, however, believed that every scrap of evidence, however provisional, was worth communicating to farmers, in part to keep them abreast of the latest work and in part to prepare them to accept changes in the company's policy or line of goods. It was also good press. Funk and his staff worked hard to keep the company name in the limelight, and there was no better way to do so than to keep reporting "new findings" that farmers would equate with prospects for increased productivity.[32]

Breeding for Resistance

The motive force behind the establishment of a federal field station at Funk Brothers was Funk's conviction that corn

[31]Funk to Woodbury, 3 October 1917, Funk Files.

[32]Funk's letter to Kellerman has not been located, but see Kellerman to Funk, 10 December 1917, Funk Files.

disease had become alarmingly prevalent in Illinois and other states and was taking a significant toll on the corn crop. Worse, no one seemed to be studying the problem from a practical point of view. By the time Funk approached the USDA he had accumulated persuasive experimental evidence that the disease problem was big and getting bigger and, more important, that his company was in an ideal position to study corn disease and develop solutions to the problem by means of selection and breeding.

Funk Brothers' research on corn disease began in 1913 during Holbert's first summer on the company farms. As discussed earlier, germination tests were used to determine the viability of selected seed. If seed showed promising germination, it was considered a good candidate for the breeding plots. The correlation between laboratory and field tests was straightforward: ignoring the scorecard ideals, ears were selected from the field on the basis of their healthy appearance, and this judgment was tested further with the germinator. Those ears that evinced poor viability or disease on the germinator were destroyed; those that did well were planted. As Hackleman later put it, the germinator was "the acid test."[33]

After studying this correlation for a few years, however, Funk concluded that it was far from foolproof. Seed that had passed the germinator test often produced diseased corn in the field, while seed that seemed weak in the germinator often produced healthy field corn. Funk's assumption that like begets like—healthy seed produces healthy corn—appeared erroneous, but how good seed became diseased remained a mystery.[34]

In 1917 Funk noticed an alarming increase in the amount of seed his assistants were discarding as diseased. In the germination room, tray after tray of seed from apparently healthy ears was thrown out because it had germinated poorly. On checking the germinator itself, one assistant discovered that when an earlier

[33]J. C. Hackleman, "Annual Report of the Crops Extension Specialist for 1931" (Office of Crops Extension Specialist, University of Illinois, Department of Agronomy), p. 15.

[34]Holbert, "Germination Story" (manuscript, 1948, Funk Files), p. 4.

batch of moldy seedlings had been thrown out, a few moldy seeds had remained in the machine, and every batch of seed that followed was thus infected. When the machine was taken apart and rebuilt with fresh material, the disease problem vanished. Neither Funk nor his staff had realized that disease could spread so easily to otherwise healthy seed and, what was worse, no one expected the germinator to be so vulnerable.[35]

Funk and Holbert took this information to Hoffer at Purdue, and by 1917 both research groups had begun studying the problem systematically under the auspices of the USDA. Hoffer discovered that several discrete organisms were responsible for the problem and that disease could be transmitted in the seed from one generation to the next. Often the disease did not interfere with seed germination but appeared only after plant growth was established. In their first national report on the situation, Holbert and Hoffer claimed that "the rate of seedling development usually referred to as 'vitality'" was not an indication of "freedom from infestation of the seed by bacteria and a species of Fusarium," nor was it "indicative of the yield possibilities of that seed ear." In other words, the germinator was an unreliable guide in selecting field seed.[36]

Despite the germinator's poor showing, of course, it was not

[35]Funk, "The Search for Better Corn," pp. 6–7; FBSC, "History of Hybrid Corn," pp. 14–16.

[36]In his *Professor's Story of Hybrid Corn* (p. 32), Hayes briefly discusses Hoffer's work but does not relate it to the Funk Brothers project. See also Benjamin Wallace Douglass, "A Revolution in Corn Growing," *Country Gentleman*, 1920, 5, 47; G. N. Hoffer and J. R. Holbert, "Results of Corn Disease Investigations," *Science*, 1918, 47:246–247; G. N. Hoffer and J. R. Holbert, "Selection of Disease-Free Seed Corn," *Indiana AES Bulletin* 244, 1918. Douglass's article, which credits Hoffer with pioneering the study of corn disease, was criticized by an unnamed Illinois farm adviser (almost certainly D. O. Thompson), who countered that Hoffer got involved years after Funk brought the problem to his attention; see Farm Advisor [*sic*] to Barton W. Currie, editor, *Country Gentleman*, 30 January 1920 (copy in Funk Files). This article also rankled Holbert, who wrote to C. R. Ball, "When reference is made to other parties concerned we should insist on sticking to the facts. Articles of this nature are not very conducive to a spirit of cooperation" (Holbert to Ball, 26 January 1920, records OCI, 54/31/61, NA).

feasible to eliminate its use altogether. Rather, Holbert and his colleagues effected a compromise and in a 1920 paper for the *USDA Farmers' Bulletin*, described how to build and operate a germinator properly. While they did not communicate their earlier findings on germinator reliability, they did point out that "great care is necessary in reading results on the germinator. Some seedlings which appear healthy grow from kernels rotted on the inside. The only safe practice is to cut open lengthwise all kernels not evidently rotted." This technique, called the "jack-knife method" by DeKalb breeder Charles Gunn, was familiar to corn show judges, who considered germination an important factor in the scorecard tally.[37]

At Funk Brothers, germination tests continued to serve as the primary means of detecting corn disease. Throughout the 1920s the company graphically illustrated the differences between "poor seed" and "good seed" in their catalogues, linking good seed to good germination. In 1922 Funk complained that "we are running our germinators to capacity and throwing out from 25 to 50% of diseased infected corn." In addition to offering "seed corn germinated for vigor and freedom from disease" at the rate of twelve dollars a bushel, or more than double the usual price, Funk also offered seed testing at three cents an ear for five hundred ears.[38]

Eager to get these new findings across to farmers, Holbert and, through him, Funk Brothers cooperated with the University of Illinois extension service in developing a set of guidelines for farmers to follow in obtaining good seed corn. In their 1920 paper Holbert and Hoffer described the visual manifestations of diseased corn in the laboratory and field, noting the symptoms in seedlings, young plants, and mature plants. In a report to colleagues they also admitted the difficulty in identifying diseases

[37]J. R. Holbert and G. N. Hoffer, "Control of the Root, Stalk, and Ear Rot Diseases of Corn," *USDA Farmers Bulletin* 1176, 1920, 19; Charles Gunn, interview with Don Duncan, DeKalb Files.

[38]See Funk to George Stevenson, 28 March 1922, quoted in Cavenagh, *Seed, Soil, and Science*, p. 268; FBSC, 1921 catalogue, pp. 2–5, 1925 catalogue, p. 6.

because "the symptoms so closely resemble and are so intimately related to the effects of drought, low fertility of soil, poor drainage, and insect and frost injuries." In both reports they concluded that until resistant strains could be developed, "the most successful control measure that has been developed so far is the selection of seed from healthy, vigorous plants that show no evidence of disease upon germination, culling out by physical selection those ear types that have been found to be more susceptible to these troubles."[39]

When, by 1922, the college was organized both for pathological investigations and extension work, Holbert was able to circulate his findings through the farming community by means of Hackleman's extension network. In a joint project titled "Corn Root Rot," Hackleman and Holbert tried to persuade farmers to select disease-free seed corn and test its germination, demonstrating through scattered field trials the yield of diseased corn compared with the yield of healthy corn.[40] Holbert also took advantage of the college's interest in physiological investigations, deciding in 1922 to combine his practical and theoretical interests in Ph.D. work with W. L. Burlison and Charles Hottes. His focus, not surprisingly, was the "physiological comparison of some representative inbred strains and their first-generation hybrids."[41]

By 1926 the disease studies had become both more straightforward and more complex. The inbreeding Holbert started in 1919 clearly demonstrated that inbreds varied dramatically in their susceptibility to particular corn diseases. Planting different inbreds in alternate rows, Holbert was struck by the contrast: one row would be almost entirely devastated by disease, but the next would be perfectly normal. It was a startling affirmation of the

[39]Holbert and Hoffer, "Control of Root, Stalk, and Ear Rot Diseases," pp. 6–13; Holbert and Hoffer, "Corn Root and Stalk Rots," *Phytopathology*, 1920, 10:55; Holbert, "Control of Corn Rots by Seed Selection," *Illinois AES Circular* 243, 1920.

[40]J. R. Hackleman, "Annual Report for 1923," pp. 28–31.

[41]Holbert to C. R. Ball, 28, 4, and 18 February and 15 May 1922, Records OCI, 54/31/61, NA.

range of variations represented. But the problem was complicated by the various field conditions that affected a plant's natural susceptibility to disease. For instance, the fungus *Diplodia* did more damage in cool, "old" soil than in warm, "new" soil. In some respects the worst-case scenario was apparent in 1935, when Holbert reported special vulnerability to *Diplodia* among plants injured by leaf blight, partly defoliated (as, for instance, during a hailstorm), or attacked by chinch bugs, and among those that were susceptible to cool temperatures. While Holbert was indeed discovering a number of interesting things about corn diseases, the complicating factors seemed to increase beyond comprehension.[42]

Clearly, disease resistance was but one class of variables Holbert had identified in trying to isolate healthy inbreds. Between 1926 and 1933 he studied the effect of cold temperatures on seed corn, both identifying strains resistant to cold and relating cold resistance to disease resistance (or susceptibility). This work began accidentally when corn seedlings that had been in the warm germinating room became crowded and some were moved into a much colder room. Discovering that some seedlings were destroyed by the cold while others were unharmed, Holbert launched a series of experiments designed to isolate the hardiest seeds. Cold-resistance was important in the event of unusually early fall frosts and also had the potential to expand Funk Brothers' "territory" farther north, where cold temperatures were the norm. In cooperation with Burlison at the college of agriculture and J. G. Dickson from the University of Wisconsin, Holbert constructed mobile refrigeration chambers that could be used for artificially subjecting field plants to cold temperatures.[43]

[42]A. L. Smith and Holbert, "Corn Stalk Rot and Ear Rot," *Phytopathology*, 1931, 21:129; Holbert et al., "Some Factors Affecting Infection with and Spread of *Diplodia* in the Host Tissue," *Phytopathology*, 1935, 25:1113–1114.

[43]Crabb, pp. 128–130; Holbert, "Germination Story," pp. 6–7; Cavenagh, *Seed, Soil, and Science*, pp. 394–395, 398; J. G. Dickson and Holbert, "Relation of Temperature to the Development of Disease in Plants," *American Naturalist*, 1928, 62:311–333.

Beginning in 1929, Holbert turned his attention to the third major corn threat—insects. Like his other projects, this one started accidentally. W. P. Flint, an entomologist at the University of Illinois, predicted that Illinois cornfields would be ravaged that summer by chinch bugs. Holbert was especially concerned because he was developing promising inbreds, and if they were destroyed by chinch bugs, his research program would be set back several years. After consulting with Flint, he decided to try to hide the inbreds from the bugs by planting them in the middle of a field bordered with open-pollinated strains. Nonetheless, the bugs found the inbreds, but as the season progressed Holbert was surprised to find that the bugs were selective in their destruction; while some inbreds were destroyed, others were virtually untouched. Holbert knew from experience that the apparently natural resistance of some inbreds to chinch bugs was his most potent weapon against them—and presumably other pests.[44]

The problems Holbert faced in breeding insect-resistant strains were very similar to those he found in trying to breed for disease and temperature resistance. The first problem, which had been apparent even with open-pollinated corn, was that high yield and resistance were rarely combined in the same strain. In Illinois, for example, where chinch bugs were an intermittent menace, farmers could plant the highly resistant open-pollinated Champion White Pearl, or they could plant a high-yielding strain and hope the bugs would not strike. The second problem was that resistance to one insect did not indicate resistance to others. Increased resistance to chinch bugs might create increased susceptibility to corn borers, or drought, or *Diplodia*. Breeders were staggered by the number of choices they needed to make in producing improved hybrid strains. As Holbert said, "Such considerations have forced Funk . . . corn breeders to concentrate efforts

[44]The insect studies have not been as well treated in the literature as Holbert's other projects. See Cavenagh, *Seed, Soil, and Science*, pp. 393–394, 396; Holbert, "How Corn Breeders Put Insect Resistance into Modern Hybrid Corn," 1946, pp. 27–28, Funk Files.

on the most important insect pests for the best hybrids adapted to a given region." Like any highly specialized organism, these inbreds were both very resistant to some things and very vulnerable to others, and breeding commercial hybrids was therefore an extremely complex phenomenon.[45]

These studies with insects, disease, and temperature convinced Holbert that identifying and breeding resistant strains was the most effective means of addressing chronic problems of corn farmers. He felt that breeding was more efficient than selection, which promised only gradual, generalized improvement. Breeding was also better than chemical applications, which might help if the corn was at risk but were wasted if the threat failed to materialize. Second-guessing the weather and potential diseases and pests was, of course, a perennial necessity for farmers, but if Holbert's breeding succeeded, it could be made a much less risky operation.[46]

Selling Science

Although hybrid corn did not constitute a significant part of the corn acreage in Illinois until the mid-1930s, its commercial potential was clear by 1925. In that year Lester Pfister of El Paso, Illinois, and Charles Gunn of DeKalb began breeding work that ten years later would make them serious competitors of Funk Brothers. In 1926 Henry Wallace established his Hi-Bred Corn Company, later renamed Pioneer, in Des Moines, and the following year Funk Brothers organized a hybrid corn division. With

[45]Ibid., pp. 10–11.

[46]It should be noted, however, that Funk Brothers cooperated with DuPont in the development of a seed corn disinfectant, Semesan Jr., described as "exceptionally effective in controlling seedborne infections" of *Diplodia* and other diseases (FBSC, 1928 catalogue, p. 10). See also Cavenagh, *Seed, Soil, and Science*, p. 396.

these developments and with Funk Brothers' considerable accumulation of experimental findings, the company began to consider its commercial opportunities in a new light.[47]

The first issue for Funk Brothers was creating a demand for the new hybrids. Despite the considerable virtues of hybrid corn, there were also a few drawbacks from the farmers' point of view. As Wallace noted in 1925, "The investment must be made afresh each year because this cross of inbred seed will give its unusual yield only the one year. Of course a return of five or ten bushels is very good for an outlay of 50 cents or a dollar, but it will doubtless take a lot of educational work before many farmers will care to buy the new type of seed corn." Indeed, as noted earlier, farmers tended to follow tradition, first, by resisting the innovation, and then, by trying to save hybrid seed for planting.[48]

Funk Brothers began a propaganda campaign in 1926. In a series of interviews with Curtis Bill of the Bloomington *Pantagraph*, Holbert attempted to introduce hybrids to farmers and to explain the difference between them and ordinary open-pollinates. In these interviews, which were peculiar by any standard, Holbert discussed corn in psychological terms. Thus, in 1927 he tried to explain the frequent linkage of positive and negative characteristics within one inbred in a piece titled "Traits Found Linked in Corn and Human Nature." In another piece called "Corn Doesn't Grow? Parents Are to Blame," Holbert formulated the memorable adage, "Education is to man what manure is to the pea." While it is unclear whether farmers were more intrigued or perplexed by these interviews, the pieces almost certainly let farmers know that something new was in the air.[49]

Between 1929 and 1932 Funk Brothers also offered special deals to induce skeptical farmers to try the new hybrids. Farmers could buy a package comprising enough seed of two open-polli-

[47]Leon Steele, "The Hybrid Corn Industry in the United States," in David B. Walden, ed., *Maize Breeding and Genetics* (1978), pp. 29–40.

[48]Henry A. Wallace, "The Revolution in Corn Breeding," *Prairie Farmer*, 21 March 1925, 1.

[49]Curtis Bill, *Pantagraph Daily Bulletin*, 1926 and 1927, passim.

nates and two or three hybrids to plant a quarter acre of each. In 1931 members of the 4-H Club received an extra hybrid at no cost. Farmers could also obtain an extra half bushel of free hybrid seed when they ordered five or more bushels of any sort and an extra 10 bushels with hundred-bushel orders. The catalogue explained that the company offered the free hybrid seed "in order to encourage its use on larger acreages." The most interesting inducement of all, however, was explained in an advertisement labeled "Win a Prize—Watch It Grow": "Send us the names and addresses of ten or more good farmers and we will send you enough hybrid corn for ten hills free. We will give $10 for the best yield from this seed and twenty prizes of $1 for the next twenty highest. Only boys and girls under 20 years of age are eligible. A day will be set aside to entertain all contestants at Funk Farms Corn Breeding Plot next September."[50]

The methods of persuasion were not new; rather, they combined the most successful features of extension service strategies. The advertisement addressed to boys and girls was a patent imitation of the extension use of Boys' Clubs and Girls' Clubs to shame parents into adopting new techniques. The seed giveaway, long practiced commercially when a company introduced a new product, had also served as a form of farm demonstration. Extension agents had early found that the most provocative form of persuasion consisted in showing ordinary farmers how more adventurous farmers increased their crop yield by adopting new practices. Funk Brothers was certain that the hybrid seed sample would convince the farmers planting the seed of the superiority of hybrids, and would convince their neighbors as well. Finally, this strategy held a further advantage that again was familiar to extension agents: every farmer who planted a hybrid sample, for better or worse, would be providing the company with useful information on that strain's range of adaptability. If it did not do well in a particular area or under particular conditions, the company was sure to hear about it.

[50]FBSC catalogues for 1929, p. 16; 1930, p. 8; 1931, pp. 5–6, 9; 1932, p. 17.

Nonetheless, sales of hybrid seed were negligible until 1933, when the company started keeping records on the subject. In that year they sold just over 810 bushels—enough to plant about sixty-five hundred acres—at the average rate of $4.11 per bushel. That year the company sold over ten times more open-pollinated corn, at the much lower rate of $1.48 per bushel. Interestingly, the price of both hybrids and open-pollinates climbed steadily, peaking at $9.46 for hybrids and $3.55 for open-pollinates in 1937, when farmers switched in staggering numbers from open-pollinates to hybrid and by which time Funk Brothers no longer led the market in sales.[51]

Dividing the Pie

Throughout the 1920s and early 1930s the relationship between Funk Brothers and the University of Illinois remained cordial but rather distant. Holbert published scientific reports with members of the Agronomy Department and in fact was first author in many experiment station reports. He also cooperated with Hackleman, and helped devise many extension strategies and projects. Eugene Funk continued to serve on the Farm Crops Advisory Committee, where, because of his experience in both dealing with the college and conducting crop research, he presumably exerted considerable influence. The only detectable source of conflict during this period related to the Illinois Crop Improvement Association. In 1934 Funk Brothers decided not to certify its corn, both because of the cost and, most important, because the company felt it did not need a recommendation from the ICIA to sell its seed successfully. This development should have sounded alarm bells at the university because it represented a distinct challenge to university control over the production and

[51]FBSC, "Application for Relief," exhibit 10.

marketing of hybrids in Illinois. This issue did not develop into a crisis, however, until after 1936, when hybrids had achieved wide prominence.[52]

In the intervening years a more pressing concern for Funk Brothers was the peculiar relationship between scientific knowledge and business practice. Holbert and the company had arrived at an understanding about their relations to each other, but the success of Holbert's research obliged them to clarify the situation. Put another way, the relationship between the company and the USDA was unproblematic when the efforts of both were directed toward investigations of disease and pest resistance. Once the product of these studies—hybrid corn—became an economic reality, once it was transformed from a research topic to an agricultural commodity, the relationship changed. It was as if the hat Holbert and Funk had both worn were traded in on two new hats—a company hat for Funk and a USDA hat for Holbert. Thus, a breeder requesting inbreds from Funk would likely be rebuffed, as good business practice would dictate, whereas one asking the same from Holbert would be obliged, as government policy dictated.

As a company, Funk Brothers did not "share" seed with other growers. When, in 1925, the Henry Field Seed Company wanted to purchase an inbred strain Funk Brothers had developed, Funk declined on the grounds that because plant breeders have no patent, they must retain the original strain as protection. He went on, "If there is anything to this work of crossing or line breeding, . . . we can hardly be expected to pass these 'pure line strains' along over to our best friends and the writer considers Henry Field one of them." The main danger, to Funk's mind, was that the recipient would simply sell the seed, unchanged, under his own name, as Funk alleged had been done in the past. But while this

[52]C. M. Woodworth to J. C. Hackleman, 4 September 1934, 8/6/2, box 9, UIA. The growing conflict between the seed companies and the ICIA will be discussed in Chapter 6.

was an understandable concern with open-pollinates, it was inappropriate with hybrids. An inbred strain was not salable until it was crossed with another inbred with which it was compatible and which demonstrated superiority over available open-pollinates. That is, unless the breeder receiving the inbred or single-cross seed had something to cross it with, and the knowledge and facilities to do it properly, the inbred alone was of little value.[53]

Rather than share or even sell to other seed producers, Funk Brothers created a network of associate growers, through which smaller corn producers were effectively co-opted by the company. The objects of the associate-growers plan were to extend the range of conditions under which Funk Brothers' hybrids were grown and to enlarge its market by scattering agents throughout the corn-growing region. The program started in 1932 when Funk Brothers supplied Clare V. Golden with enough of two strains of single-cross seed to produce twenty acres of hybrid corn. Golden then sold the hybrid seed himself, paying Funk Brothers a royalty of 15 percent of his net sales. The number of associates increased each year; by the mid-1950s there were seventeen scattered around the country, with the highest concentration in the Midwest.[54]

The associate-growers strategy served another important function, that of absorbing the competition. Most of the associates had been prominent seed producers in their own right before the advent of hybrid corn and had competed with Funk Brothers in the production and sale of open-pollinated corn and other seeds. Some, such as McKeighan Seed Company in Yates City, Illinois, J. C. Robinson Seed Company in Waterloo, Nebraska, Sommer Brothers Seed Company in Pekin, Illinois, and A. H. Hoffman Seeds in Landisville, Pennsylvania, had been in business twenty-five or thirty years. Few, if any, of them, however, had the resources or expertise to do the costly and extensive research nec-

[53]Funk to Henry Field, 24 March 1925, Funk Files.

[54]FBSC, "Application for Relief," pp. 60–65; Cavenagh, *Seed, Soil, and Science*, appendix V, "Funk's G-Hybrid Associate Producers," pp. 470–471.

essary to develop inbred strains with which to create hybrids. As a company report noted, these seed houses faced a difficult choice: "either convert to hybrid seed corn by securing this item for sale or go out of the seed corn business." Funk Brothers liked to think that the associates plan "provided an avenue whereby these producers might continue in business," although it is hard to imagine that these formerly independent producers felt only gratitude toward the company.[55]

The larger seed producers, such as Pioneer, DeKalb, and Pfister, benefited not from Funk Brothers but from Holbert, who was generous in distributing his inbreds and single crosses. In 1922 E. H. Jenkins at Iowa State University obtained one of Holbert's first single crosses, which he crossed with an inbred of Iodent that Henry Wallace had been working on. In 1932 Henry Wallace bought single-cross seed from Holbert for crossing with his own lines. Pfister and DeKalb also received lines from Holbert in the 1930s. This open exchange of breeding stock between the experiment stations and the seed companies apparently ended in 1937 when the federal field station at Funk Brothers was closed.[56]

The Successful Merger

It would be simplistic to conclude that the success of Funk Brothers resulted from the federal government's largesse. More important was the role of Eugene Funk in devising agendas for corn research and in identifying and enlisting promising scien-

[55]FBSC, "Application for Relief."

[56]Henry Wallace, "Public and Private Contributions to Hybrid Corn, Past and Future," *Proceedings of the 10th Annual Hybrid Corn Industry Research Conference* (Washington, D.C.: American Seed Trade Association, 1955), p. 4; Holbert, "Testimony Offered in the Case of James Arleigh Batson vs. J. C. Robinson Seed Company," 24 and 25 September 1956, Funk Files, pp. 76–77; Wallace, "Revolution in Corn Breeding," pp. 1, 15; Wallace to Funk, 3 March 1932, Funk Files; Crabb, pp. 189–190.

tists such as Holbert. Unquestionably Holbert's resistance studies played a critical role in the company's early development of hybrids. Yet it should be noted as well that one of the most eminent corn breeders in the country advised Holbert that working with Funk would be Holbert's best professional opportunity. It is a moot point whether Funk Brothers would have acquired the federal field station had Holbert not been in residence or whether Holbert would have opted to leave Funk Brothers had that been a condition of his appointment to a station located elsewhere.

Clearly, Funk Brothers practiced a kind of corn research and extension fundamentally different from the pattern at the University of Illinois College of Agriculture. First, the company treated corn as a commodity rather than an experimental subject. College breeders were interested in studying how morphological and physiological characters were inherited; sometimes, though not very often, they focused on economically significant characters. Funk Brothers, on the other hand, while attentive to Mendelism, was far more interested in what Mendel's methods could do for corn production than in what they could reveal about patterns of inheritance.

A second difference lay in the kind of attention each group could devote to corn studies. Funk Brothers conducted research on and marketed a wide variety of other seeds and grains, but its most abiding interest was corn. In terms of acreage, staff, and financial investment, the company spared little in its efforts to increase the yield of corn. The college, however, faced constraints at all levels. Because its landholdings were small, the extent of field research was much more limited, and researchers often relied on cooperating farmers to make up the difference. The college budget, controlled by all manner of outside parties, including the board of regents, the state legislature, and the federal government, was constantly inadequate. Perhaps more important, the corn work was but one aspect of agricultural research and development at the station; as pointed out earlier, projects such as the soil survey absorbed a great share of the resources.[57]

[57]Hopkins's oil and protein selection experiments might appear oriented toward economic considerations, but Hopkins—a chemist rather than a breeder—

As I turn in the next chapter to an examination of corn development at DeKalb and Pfister, the roles of Funk Brothers and the university will become even clearer. Just as the assets and liabilities of Funk Brothers and the university differed, so the available resources and constraints at these two companies were in many ways peculiar. It is necessary to understand the range of commercial experience before the similarities, differences, and interrelationships between public and private agricultural research can be illuminated.

---

was far more interested in, if not fascinated by, the sheer manipulation of molecules than in the practical potential of his work. That the experiments continued for more than a decade after the commercial applications failed would seem to support this view.

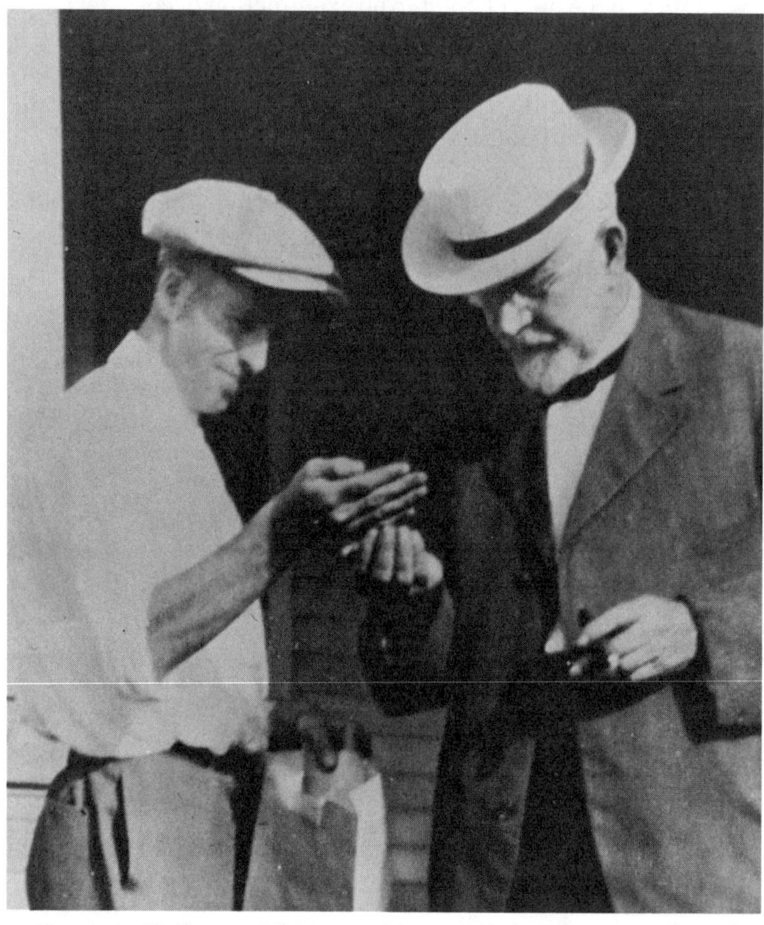

Henry A. Wallace exhibiting to his grandfather Henry Wallace the results of cross-fertilizing Boone County White and Early Wisconsin corn in 1913. (Photo courtesy of the State Historical Society of Iowa, Special Collections.)

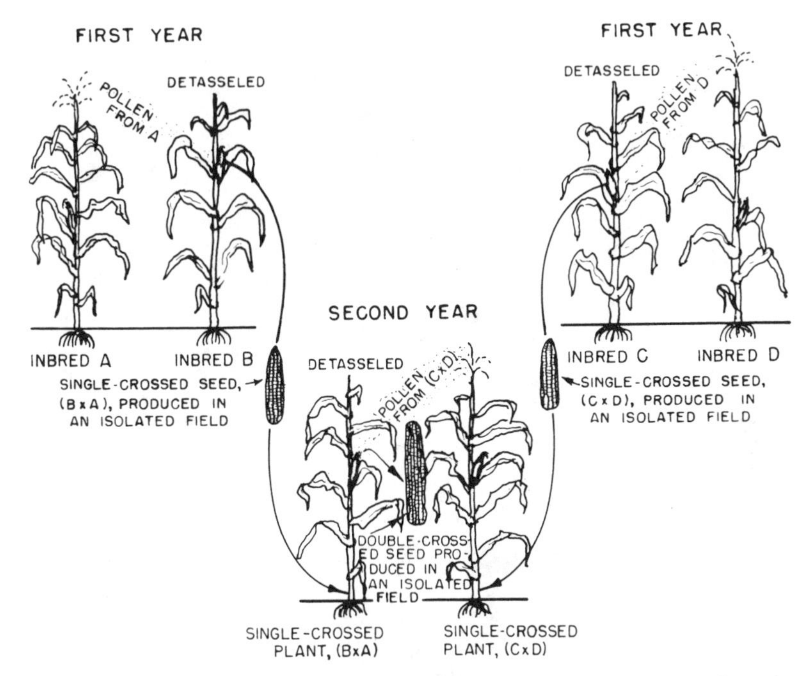

The double-cross method of hybrid seed production. (Reprinted, with permission, from Herbert Kendall Hayes, *Professor's Story of Hybrid Corn* [Minneapolis: Burgess, 1963].)

Frederick D. Richey, the most influential USDA breeder. (Reprinted, with permission, from Hayes, *Professor's Story of Hybrid Corn*.)

Donald F. Jones, developer of double-cross hybrids, and Henry A. Wallace in 1950. (Reprinted, with permission, from Hayes, *Professor's Story of Hybrid Corn.*)

Eugene Davenport, dean of the University of Illinois College of Agriculture. (Photo courtesy of the University of Illinois Archives.)

J. C. Hackleman, founder of the Illinois Crop Improvement Association and first director of the Division of Crops Extension at the University of Illinois. (Photo courtesy of the University of Illinois Archives.)

Eugene D. Funk, the scientifically inclined seedsman whose company, Funk Brothers, was in the forefront of hybrid corn development. (Photo from the Funk Seeds International files, reproduced courtesy of Leon Steele.)

(Top right) J. R. Holbert, director of the federal field station at the Funk Brothers Seed Company farm. (Photo from the Funk Seeds International files, reproduced courtesy of Leon Steele.)

(Bottom right) Lester Pfister, self-taught breeder and founder of Pfister Hybrid Corn Company. (Photo from the Pfister Hybrid Corn Company files, reproduced courtesy of Daniel E. Pfister.)

DeKalb Agricultural Association's "Winged Ear" is depicted in an early advertisement touting the drought resistance of hybrid corn. (Courtesy of DeKalb-Pfizer Genetics.)

6

A New Game:

Adjusting to Hybrid Corn

Between 1936 and 1942 the excitement over hybrid corn in Illinois created an uncertainty in the College of Agriculture over its role in developing, distributing, and regulating hybrids. It was becoming clear to college officials that as the value of hybrids shifted from the sphere of experiment and demonstration to the more thoroughly commercial, the college's relationship with both seed producers and farmers would also change. The issues raised by the commercialization of hybrids were complex and unprecedented, and the college and seed producers were increasingly divided over how to resolve them. As the seed companies shifted from the experimental arena, in which cooperation with college breeders as well as other private breeders worked to everyone's advantage, to the commercial arena, where cooperation was replaced by competition and secrecy, the college had to shift its ground. It was a noncommercial enterprise producing a commercial product, hybrid corn. The questions now were what to do with these hybrids and, perhaps more important, what to do with the commercial producers.

At issue was the commitment to public service. In the early days the agricultural college fulfilled its public service obligations by making the experiment station staff available to farmers, helping farmers organize Farm Bureaus and other agricultural associations, and trying to devise research agenda with the farmers' interests in mind. The college had always walked a fine line in these matters, attempting to distinguish its educational activities from their commercial consequences. The public service had also extended to the commercial community at times, and throughout the long period of experimentation on corn and hybridization, the college happily cooperated with seed producers in the name of science. As officials often pointed out, the college could not discriminate against any segment of the farm population to which the scientific authority and expertise of the college would be of value; seed producers were people too, so the logic went, and the college, in its role as advocate for the farm population, was obliged to aid and advise anyone who had the wit to ask for help.

In effect, the commercialization of hybrids changed the rules of the game. The so-called farm population had never in fact been as homogeneous as some college policy suggested, and this new development seemed to accentuate the differences. For example, where before there had been one large seed house (Funk Brothers) and perhaps a hundred smaller ones, by 1936 there were three large seed houses competing vigorously with each other as well as with small producers. And as the structure of the game changed, so did the abuses of rules. Irresponsible or ignorant producers were dumping unadapted hybrid seed on the market or selling second-generation hybrids. Breeders who had once traded inbred and single-cross lines now changed their original hybrid designation, hopelessly tangling the pedigrees of many lines. Large producers began balking at certification, threatening to destroy the "good faith" understanding upon which the certification program was built.

DeKalb Agricultural Association

By the mid-1930s the hybrid corn industry in Illinois was dominated by Funk Brothers, Pfister, and the DeKalb Agricultural Association, no one of which could be considered "representative" of the industry as a whole. Despite fundamental and necessary similarities—all three developed inbred lines of corn, crossed them, and sold the resulting hybrids—they were quite different in their origins and historical development, and one searching for a commercial model upon which to base a new company would have been hard pressed to distill the essential components from these disparate entities. Yet together, these companies brought about the rapid commercialization of hybrid corn in Illinois in the 1930s and the concomitant challenge to college authority and control after 1935, when DeKalb and Pfister joined Funk in marketing the hybrids.[1]

In early 1912, a dozen affluent farmers in DeKalb County, Illinois, got together and decided that they needed to acquire an agricultural expert who would help them improve their farming practice. This sentiment and the entrepreneurial spirit behind it represented the melding of two recent agricultural innovations: the county agent system and the regional drive to improve Illinois soil. By 1912 the county agent system, inaugurated in the southern United States in 1904, was beginning to receive positive attention in the North. Many farmers who had depended upon the loosely organized Farmers' Institutes felt that a county agent would be a more efficient and useful channel through which to receive practical advice and agricultural news. Although the college employed Fred Rankin as an extension agent, Rankin was expected to serve the entire state and did not have the time or resources to devote much attention to DeKalb County. In addition, DeKalb County farmers, who lived on the northern edge of the state, felt both geographically and psychologically isolated from the University of Illinois, center of agricultural expertise in the state.[2]

Moreover, beginning in 1902, agronomists at the college began

[1]The fourth major hybrid seed producer—Pioneer Hi-Bred Corn—did not face the same degree of competition in Iowa that Illinois companies contended with; Pioneer's two rivals—Michael-Leonard in Sioux City and National in Anamosa—were not as influential as the "big four."

[2]John Lacey, *Farm Bureau in Illinois*, pp. 11–13.

"mapping" the soils of Illinois, that is, developing a physical description and chemical inventory of Illinois soils. The first soil survey report, which appeared in 1911, persuaded many farmers that their own farmland was deficient in essential nutrients. C. G. Hopkins launched a popular crusade across the state to alert not only farmers but educators, bankers, and merchants that Illinois soil fertility was on the decline but could be restored by the judicious application of fertilizer. Speaking before the Illinois Bankers' Association, Hopkins described the situation in language bankers could appreciate: "Consider the soil as a bank in which we may have a fair bank account and can withdraw from it only a small percentage each year of what it contains. We cannot withdraw from this bank the total deposit at any one time. We can withdraw one or two percent at a time, because there is a law of nature relating to the soil which makes it impossible for us to withdraw from the soil at any one time all that is in it. It does not become available; it cannot be made available."[3]

It was one thing to know that the soil was deficient; it was quite another to prescribe the proper chemical treatment. The DeKalb group wanted to procure not just any agricultural expert but a "soil doctor," who would examine each farmer's land and recommend appropriate remedies. W. G. Eckhardt, a soils expert from the college and protégé of Hopkins, was hired by the DeKalb County Board of Supervisors soon after the farmers' group formed. Like those of the college experiment station at the turn of the century, Eckhardt's services were in great demand from the start. His primary job was to visit the farms of all interested

[3]C. G. Hopkins, "The Soil as a Bank" (address to the Fourth Annual Convention of the Illinois Bankers' Association, 15 June 1910, UIA), pp. 3–4. Hopkins was evangelistic on the subject of soil fertility; see "The Story of a Corn King and Queen," *Popular Science Monthly*, March 1911, 251–257, a moral tale in which the failing health and vitality of King Corn and Queen Clover are put right by Doctor Science, whose "knowledge was true and absolute." Dr. Science, who prescribed a diet of limestone and phosphorus and recommended rotation, represented Hopkins himself. See also University of Illinois, Department of Agronomy, "Historical Data for President Kinley," February 1941, 8/6/2, UIA, pp. 63–72.

members of the association, assess the crop and soil condition, and suggest how the existing situation could be improved. He was also put in charge of the county demonstration farm, which he turned to good use by providing demonstration plots that reflected recent soil and crop recommendations.[4]

It soon became clear to these enterprising farmers that soil was only part of their problem. Seed that was properly adapted to the area was also in short supply, and poor seed threatened to negate their soil-improvement efforts. Located on the far northern fringe of the corn belt, DeKalb County farmers had always found it difficult to obtain field seed that matured soon enough to avoid the early frosts of autumn. Most corn and field seed, bred in the more moderate climatic zones of central Illinois, was gradually and imperfectly adapted to the north only after repeated planting.[5]

In 1917 the Soil Improvement Association became the DeKalb County Agricultural Association. The new name signified an enlarged range of activities, including the procurement of seed adapted to the DeKalb growing season. Eckhardt, who retained leadership of the group, felt that better field seed could be obtained from growers and breeders in northwestern states such as Idaho, where climatic conditions were even more severe. With this idea in mind, he persuaded Charles Gunn to join the association in 1917 as "seed scout."[6]

A thirty-one-year-old native of rural Illinois, Charlie Gunn had accumulated considerable practical experience in agriculture before joining the DeKalb group. He grew up on a farm, and in 1905, at the age of nineteen, began working for Walter Griffith, a reputable seedsman in the area. Griffith, like his friend Eugene Funk, took an early interest in corn breeding and was a member of both the Illinois Corn Breeders' and the American Breeders' associa-

[4]Lacey, *Farm Bureau in Illinois*, pp. 13–20.

[5]DeKalb Agricultural Association, "The Twelve Apostles of Farm Cooperation," *Winged Ear*, 1939, 1, DeKalb Files.

[6]Crabb, *Hybrid Corn Makers*, pp. 210–228.

tions. Like many of the seedsmen newly acquainted with breeding techniques, Griffith was eager to experiment with the latest methods even when he did not understand them. As Gunn recalled, "I can very distinctly remember going out in Griffith's orchard and helping him detassel corn. We didn't know just what we were doing." Working with Griffith, Gunn learned how to select improved corn for commercial purposes and how to check such corn for disease with the jackknife test. In 1913 Gunn moved to Colorado, where he farmed for four years, growing, among other things, corn and sweet clover. It was on a visit to Illinois that he met Eckhardt and agreed to work for the association.[7]

As seed scout, Gunn divided his efforts between procuring field seed and developing improved lines of open-pollinated corn. Each autumn he traveled to Idaho, where he bought large quantities of red clover and alfalfa seed for DeKalb-area farmers. Such legumes were an important component of the soil improvement scheme the association had devised. Eckhardt suggested the line Gunn's corn work was to follow. Eckhardt had brought a corn strain called Western Plowman into the area from southwestern Illinois. Although the strain showed promise, it was not adapted to northern Illinois, so Eckhardt assigned Gunn the task of improving and adapting it to local conditions. During the next few years Gunn selected for early maturity and finally produced a strain that was better adapted to DeKalb County. But like Funk, he discovered that there was a limit on accumulating all the favorable characteristics in one strain; as maturity approached the ideal level, yield dropped. Gunn thus became attentive in the early years to new breeding ideas.[8]

[7]Crabb, *Hybrid Corn Makers*, pp. 212–213; Charles Gunn, interview by Don Duncan, DeKalb Files, pp. 4–5.

[8]Crabb, *Hybrid Corn Makers*, p. 213; DeKalb Agricultural Association, "Sales Manual for DeKalb Quality Hybrid Seed Corn—History" (1938, DeKalb files), pp. 2–3. In 1929 a dispute erupted between Gunn and Eckhardt when both tried to certify their seed. Eckhardt sued the Farm Bureau for selling Western Plowman developed by Gunn, claiming that he (Eckhardt) was the source of the strain. The ICIA, called in to settle the matter, recognized two strains of the seed: Gunn's Western Plowman and Eckhardt's Western Plowman. See Lang, *Fifty Years of*

During the years 1917 to 1924 the social dynamic of the association was somewhat precarious. Late in the 1910s, Eckhardt left the association to take up a war-related position in Chicago. Returning to the area in 1921, he established a rival seed business that association leaders feared would divide the loyalties of their membership, and so the association reportedly began looking for attention-getting innovations that would provide a competitive advantage.[9]

The second precarious arrangement dated from the association's founding, at which time it was identified—and in fact hailed—as the first county Farm Bureau in Illinois. Indeed, the manner of its foundation was perfectly consistent with the procedure administrators considered ideal. The key components were a formal organization of local farmers who could identify their agricultural problems well enough to request assistance, the assignment of a college expert to act as an adviser to the group, and a financial commitment from the farmers to help defray the adviser's expenses. In Illinois this arrangement usually took on a commercial component as well, in that the participating farmers organized themselves into a buying cooperative for procuring bulk quantities of limestone and phosphorous needed to improve their soil according to college recommendations.

When the Soil Improvement Association spawned the DeKalb Agricultural Association in 1917, the purchase of field seed was added to its activities. This arrangement was more or less acceptable until it became clear in the mid-1920s that the DeKalb Farm Bureau and the Agricultural Association were virtually indistinguishable. College officials were particularly concerned that Tom Roberts was at once both farm adviser and secretary-treasurer of the Agricultural Association. By 1925 the association had grown from a limited buying cooperative into a distinctly commercial operation, but the Farm Bureau, college officials tirelessly

---

Service, p. 25; Hackleman, "Annual Report of the Crops Extension Specialist for 1929" (Office of the Crops Extension Specialist, University of Illinois, Department of Agronomy), pp. 35–36.

[9]Crabb, *Hybrid Corn Makers*, p. 211.

insisted, was intended to be an educational organization and the farm adviser was meant to be strictly advisory.[10]

As we have seen, it had always been difficult for the advisers to adhere to these strict definitions, and it is not surprising that personnel of the Farm Bureaus, the Illinois Agricultural Association, and the emerging farmers' cooperatives overlapped to a large extent. But the college found the situation in DeKalb County particularly intolerable, and after repeatedly and fruitlessly urging Roberts to resign one of his appointments, the college severed relations with DeKalb County in 1927. After two years the Farm Bureau asked to be reinstated, promising to do better next time, and it was allowed to return. But the following year Dean Mumford was outraged to learn that Roberts continued to serve in a dual capacity; finally in 1931 Roberts left the Farm Bureau permanently.[11]

It appears that Gunn and Roberts first considered improving corn through inbreeding in the summer of 1923, when Henry A. Wallace, son of secretary of agriculture H. C. Wallace and a "gentleman corn breeder," told them of his inbreeding work while visiting the DeKalb County Farm Bureau annual picnic. Inspired by Wallace's prediction that inbreeding and crossing corn would soon revolutionize corn production, Gunn and Roberts spent the fall and winter months consulting with corn breeders including J. R. Holbert, G. N. Hoffer, and E. H. Jenkins. These breeders had been experimenting long enough to produce a few inbred lines, which they shared with Gunn and Roberts. In 1924 these inbreds were planted in DeKalb, but their late maturity made them unsuitable to northern conditions. They seemed promising, however, and the following year Gunn started selecting and inbreeding with special attention to early maturity, which continued to distinguish the hybrid lines developed at DeKalb. Fi-

[10]See, e.g., W. H. Smith to DeKalb County Farm Bureau, 10 May 1923, 8/1/2, box 35, UIA.

[11]J. D. Bilsborrow to H. W. Mumford, 9 and 10 June 1925, 23 May 1927, 1 July 1929, Mumford to Bilsborrow, 30 April 1930, Bilsborrow to J. C. Spitler, 3 June 1931, 8/1/2, box 35, UIA.

nally, after the association's reconciliation with the college in 1931, Gunn began taking advantage of station inbreds, which he crossed with both his own and those obtained from other breeders.[12]

The exchange of inbred lines between breeders, and especially between Holbert and Gunn, continued in the mid-1930s. As a federal agent, Holbert gave Gunn access to the specialized lines he was developing in Bloomington and provided him with lines obtained from other private breeders and experiment stations. Like Gunn, Holbert was interested in developing lines particularly suited to the northern regions, and he passed on those inbreds that seemed appropriate to the task. In addition Holbert advised Gunn in his breeding work, suggesting which crosses to make or warning him against lines Holbert knew to be unsatisfactory. He helped Gunn identify unknown lines, and in general served as the regional expert on corn-breeding matters.[13]

Lester Pfister and His Company

In 1938 George Kent wrote an article that chronicled the "amazing success story" of Lester Pfister of El Paso, Illinois. Comparing Pfister to Thomas Edison and Marie Curie, Kent described his breeding work in language more romantic than realistic:

> As we hammered along the road to his farm, the story emerged, bit by bit—a tale of hardship, torment, of a long desperate strug-

[12]It seems somewhat implausible that the association had not tried inbreeding before then, since Wallace reported his research frequently in *Wallaces' Farmer* (see, e.g., issues for 7 March 1919 and 23 December 1921) and produced his first commercial single cross, called Copper Cross, in 1923. See William L. Brown, "H. A. Wallace and the Development of Hybrid Corn," *Annals of Iowa*, 1983, 47:167–179.

[13]Crabb, *Hybrid Corn Makers*, pp. 210–218; Gunn interview, pp. 34–35; Holbert to Gunn, 19 May, 18 June 1934, 22 January, 7 March, 10, 24 December 1935, in DeKalb Files.

> gle climaxed with a motion picture finish; literally a saga of rags to riches staged on the flat black earth of central Illinois. For ten years, up until 1935, some of his neighbors were convinced that he wasn't quite right in the head. They couldn't understand why any sane individual should spend hours in a field under the boiling sun tying paper bags on corn tassels. They told him so. And when his farm went to ruin because he couldn't give it the time it required, fatherly old men used to stop him on the road and beg him to quit his foolishness. The younger man only spat in the dust and grinned.[14]

And so on. The sensational accounts that continued to appear in the national press served to not only popularize Pfister and his company, but also to convince the public that corn breeding was a commonsense operation that even a poor, uneducated farmer could master.[15]

On a certain level this perception was accurate. Pfister represented the far end of a commercial spectrum that ranged from long-established seed houses with breeding experience (Funk) to soil and seed associations that employed farmer-seedsmen with little or no academic training but considerable practical knowledge (DeKalb) to independent farmers, undereducated, with sufficient but not overwhelming experience (Pfister). Yet though Pfister lacked the financial advantages, academic training, political influence, and cropland of Funk, he possessed an unusual degree of natural ability, determination, and according to Charlie Gunn, no small amount of luck.[16]

Lester Pfister was born in rural Illinois in 1897. He left school in 1911 to begin farming. Reportedly, he was attentive to the differences in yield and quality between various strains of corn, and

[14]George Kent, "Bagging a Million," *Country Home Magazine*, August 1938, 7, 21, repr. *Reader's Digest*, September 1938.

[15]See, e.g., Francis Oliver, "Success Saga of the Corn Belt," *St. Louis Post-Dispatch*, 22 September 1938, 1; Robert C. Hegleson, "Hybrid Corn Finds Market in Seed Trade," *National Seedsman*, October 1938, 8–9; and countless short pieces in the Bloomington *Pantagraph* and Chicago *Daily News*.

[16]Gunn interview, p. 36.

used ear-to-row breeding to determine which strains were most productive on his farm. In 1919 Martin L. Mosher became farm adviser in Woodford County, and Pfister availed himself of the opportunity to consult a corn expert on his own ear-to-row tests. Mosher was interested in Pfister's work, and when Mosher inaugurated a three-year yield test designed to identify the most productive corn in the county, he appointed Pfister the task of comparing and reporting on the test plots. When the test results were tallied in 1922, the strain entered by area farmer George Krug so overwhelmed the other entries that Pfister began growing open-pollinated Krug corn for sale the next season.[17]

Like Charlie Gunn, Pfister credited Henry Wallace with inspiring him to begin inbreeding work on Krug lines in 1925. He was also familiar with the work of Holbert and C. M. Woodworth, and read Wallace's reports on his own experimentation in *Wallaces' Farmer*. In 1929 Pfister began crossing his inbreds and in 1932 produced his first double-cross hybrid, Pfister's 4857.[18]

Pfister's experiments had attracted the attention of other breeders by the early 1930s. In 1930 F. D. Richey visited Pfister's farm and persuaded him to let the USDA test his most impressive inbred, number 187, in Washington. This inbred, which Pfister had self-pollinated for six generations, performed well for Richey and was used in producing US44, a USDA hybrid first distributed to experiment stations in 1935. Pfister was later angered by this use of his inbred line, maintaining that he had given it to Richey for experimental purposes only.[19]

In 1932 Pfister's breeding program also benefited from the USDA program at Funk Brothers. After seeing Pfister's inbreds, Holbert offered to test Pfister's single-cross lines in combination with his own to evaluate the combining ability of all four lines. When the University of Illinois began its corn performance tests

[17]Crabb, *Hybrid Corn Makers*, pp. 229–235; Mosher, *Early Iowa Corn Yield Tests*, pp. 82–89.

[18]Crabb, *Hybrid Corn Makers*, p. 237.

[19]Ibid., pp. 226–237; Pfister to H. P. Rusk, dean of the College of Agriculture, University of Illinois, 30 April 1941, 8/1/2, file 113, box 101, UIA.

in 1934, Holbert entered these hybrids, which a later writer called "the two greatest hybrids developed up to that time." Pfister's hybrids, numbers 360 and 366, each contained Pfister's single cross, 159 × 187, which also appeared in thirteen of the twenty-seven hybrids Pfister had available in 1938. Inbred 187, moreover, was one of four lines in all but one hybrid, and all but one hybrid contained at least one federal or station line. In other words, Pfister got a lot of mileage out of his first inbreeding efforts.[20]

Pfister's hybrid business expanded rapidly; in 1933, the first year he sold hybrids, he produced 225 bushels; four years later he produced 270,000. Pfister's greatly increased production capacity was largely the result of his Associated Grower Plan, inaugurated in 1934. Although it is disputable whether area farmers asked Pfister for the privilege of growing his hybrid seed, as popular writers reported, it is clear that the agreement was advantageous to all concerned. Pfister supplied the growers with single-cross seed, and the growers gave Pfister a 10 percent royalty on their sales of hybrid corn the next season. In this way Pfister substantially expanded the acreage planted with his corn and of course expanded his sales potential. The forty-four growers who joined Pfister by 1938 reaped the advantages of his breeding work, but some of them may have forfeited their plans to become independent seed producers as a result. Griffith Seed Company in McNabb, for instance, started early in the century by Walter Griffith, who taught Charles Gunn how to select open-pollinated corn, was one of Pfister's growers.[21]

Although many seed growers became producers for the large hybrid corn companies, that was not their only alternative. Many more were able to stay in business or, in some cases, start a

[20]Crabb, *Hybrid Corn Makers*, pp. 239–241; *Pfister Hybrid Corn News*, Fall 1938, 4, Pfister Files, lists the pedigrees of all twenty-seven hybrids then available from Pfister.

[21]*Pfister Hybrid Corn News*, Fall 1938, 2; I have not been able to determine in what year the Associated Grower Plan started. Kent dates the origin at 1932 (p. 21), but I have accepted 1934, the date given by Francis Oliver.

business, by growing and selling hybrids developed by the experiment stations. Although the college did not officially begin distributing corn lines until 1937, it is probable that they did so informally much earlier. Certainly, the large companies were able to procure station and USDA lines in the mid-1920s for combining with their own lines, and it seems likely that at least some smaller producers acquired seed as well, if only to help increase the college's supply.[22]

Adopting Hybrid Corn: Pros and Cons

By 1936 it was clear to both experiment station and commercial breeders that hybrid corn was a commodity whose time had come. Writing for the *Yearbook of Agriculture*, Merle Jenkins predicted that "the peculiarity of hybrid corn, which restricts its use to the first generation following the cross and requires that new seed be obtained each year, will of necessity result in a considerable industry for the production of this kind of seed." Farmers were also beginning to consider the potential of hybrids to boost their incomes, and although few farmers switched completely from open-pollinates to hybrids in the mid-1930s, many were willing to experiment by planting a few acres.[23]

Some scholars and contemporary observers have considered the rate at which farmers adopted hybrid corn remarkably rapid. It is important, however, to note the disadvantages of early adoption as well. The first problem, of course, was the expense. Using Funk Brothers' seed as an example, farmers paid nearly twice as much for hybrid seed as for open-pollinated in 1936, hybrid costing $7.86 per bushel and open-pollinated $3.92. The investment was good for only one year, moreover, rankling farmers accustomed to saving seed from one year to the next. Despite seed

[22]For a list of Pfister growers in 1938 see *Pfister Hybrid Corn News*, Fall 1938, 6.
[23]Merle T. Jenkins, "Corn Improvement," *Yearbook of Agriculture*, 1936, 479.

company claims that the difference would be more than offset by the increased yield from hybrids, the initial cost of seed must have seemed daunting.[24]

The second deterrent to adoption was maladaptation, which was highlighted in the yield tests conducted around Illinois by the college. In some areas of the state open-pollinates performed distinctly better than hybrids because they were adapted to local conditions, whereas the hybrids frequently seemed temperamental when removed from their original breeding ground. For instance, an open-pollinate selected over the years to resist ear worms yielded much more than a hybrid that was not designed with ear worms in mind. Open-pollinates were also superior as a rule in areas where the soil was poor or unfertilized, whereas hybrids did better on rich, highly fertilized soil. Farmers who tried hybrid corn before adapted varieties were developed were often discouraged by their poor showing and were disinclined to risk another crop failure by trying another hybrid line or brand. One farm adviser explained to Hackleman, "About ten years ago we tried four hybrids. . . . These hybrids did fine. The next year these men tried hybrids and it went down and was damaged worse than their own [open-pollinated] corn. This put a damper on hybrid corn in this county for a good many years."[25]

Adaptability was a frequent source of confusion and irritation to farmers and seedsmen alike and constituted a never-ending problem for Hackleman. As crops extension specialist, he attempted to monitor the problem by means of the annual corn performance tests, which demonstrated the relative merits of the hybrids and open-pollinates then available to farmers. Because these tests evaluated particular strains as they were or were not adapted to particular areas, they did function as a sort of sieve through which only a few of many available hybrids could pass, thus simplifying the task of selecting the best strain for a particu-

[24]Funk Brothers Seed Company, "Application for Relief" (Exhibit 10, Analysis of Net Sales, 1933–1943, Funk Files).

[25]Unidentified farm adviser to Hackleman in Hackleman, "Annual Report for 1937," p. 22, and see p. 21 and "Annual Report for 1936," p. 56.

lar farmer's conditions. But not all farmers were attentive to the tests; some assumed that hybrids were by definition superior to open-pollinates and ignored warnings to select only properly adapted lines. In addition, widespread ignorance of hybrids and the differences among different hybrid lines, as well as between hybrids and open-pollinates, paved the way for fraudulent seed sellers who found many farmers to be easy marks. Hackleman warned that "the farmer who is not familiar with hybrid corn and does not understand this variation in yield may be the innocent victim of some hybrid corn salesman who knowingly or unknowingly is selling an inferior or unadapted hybrid."[26]

Adaptation was a problem for the large seed companies as well. I do not suggest that they deliberately misled customers or purposely sold unadapted hybrids, but before 1938 there were so few hybrids in commercial production that inappropriate sales were nearly unavoidable. Just as some farmers would buy anything labeled hybrid, some loyalists would buy anything labeled Funk, Pfister, or DeKalb, regardless of suitability to local conditions. Moreover, companies seemed to feel that poor crops harvested by farmers who had planted improperly adapted lines reflected badly on their own companies and the seed industry as a whole, regardless of the source of the unadapted seed. As George Sprague later noted, the reputation of a hybrid was influenced by the poorest producer of that hybrid.[27]

The large companies could also turn the adaptation problem to their advantage. Funk Brothers, like other established seed companies, invoked its long experience in corn breeding to reassure farmers wary of hybrids and concerned that a dealer would take advantage of their lack of familiarity with this promising novelty.

[26]Hackleman, "Annual Report for 1938," p. 6. Although I have not found any specific instances of fraud, it seems likely that during the hybrid corn mania from 1936 to 1942, some considered selling supposed hybrids an ideal get-rich-quick scheme.

[27]George Sprague, "The Changing Role of the Private and Public Sectors in Corn Breeding," *35th Annual Corn-Sorghum Research Conference* (Washington, D.C.: American Seed Trade Association, 1980), pp. 2–3.

Indeed, Funk Brothers exacerbated the suspicions with warnings such as this: "Prospective purchasers must beware of seed offered as Hybrid and make sure it is good Hybrid of tested performance before they buy. The situation as it exists, large demand and limited supply, is opportune for unethical parties to throw seed corn on the market that is not truly Hybrid Seed." To be safe, in other words, the farmer should buy seed only from an established seed house such as Funk Brothers.[28]

As one might expect, Hackleman's policy under these circumstances was cautious, if not conservative. He continued to emphasize the importance of field selecting open-pollinated seed corn, the virtues of smooth over rough indentation, and so forth. He reported in 1937 that "one of the most difficult problems in connection with corn is to put on the brake," referring to the rush to hybrid corn in southern Illinois. The southern counties were distinct both in soil type and in socioeconomic characteristics. Southern Illinois soils were characteristically dense and wet, or "clayey" in local parlance, and crop yields were routinely much lower than yields in the north. Many of the farmers there were tenants rather than independent landowners, and in some counties black farmers were more numerous than white. Moreover, breeders were concentrated in the central and northern part of the state. For these reasons, no hybrids were developed for southern Illinois for many years. Some seed producers nonetheless sold unadapted hybrids to many southern farmers who, naturally, suffered poor crop yields. As late as 1938 Hackleman suggested that "farmers would do well to plant at least part of their corn to reliable open-pollinated varieties until better adapted hybrids are produced."[29]

Hackleman's efforts to restrain farmers from rushing too quickly to embrace hybrids added to the confusion surrounding adoption. Despite his repeated warnings against selecting seed from hybrid fields, it is little wonder that farmers continued to do so. By recommending that farmers continue planting open-polli-

[28]FBSC, 1936 catalogue, p. 1.

[29]Hackleman, "Annual Report for 1936," passim, "1937," p. 20, "1938," p. 23.

nates—a completely reasonable approach under the circumstances—at a time when many farmers were trying to make the transition to hybrids, Hackleman unintentionally worsened the confusion between the two. Speaking, for example, of the difficulty advisers encountered in persuading farmers that hybrid seed could not be used a second year, E. H. Jenkins recalled "one irate grower who suffered considerable financial loss because he planted some fifteen acres with seed he had harvested from his hybrid field and insisted that the warning he had received should have been vigorous enough to make him heed it whether he believed it or not."[30]

The frustration of breeders and farm advisers, whose advice often fell on deaf ears, and the perplexity of farmers, who had never before experienced such a shift in practical strategy, was also exacerbated by sensational stories about hybrids in the popular press. Hybrids became a big story in the late 1930s and early 1940s, a refreshing change from the constantly gloomy reports of the Depression. Popular fascination with hybrids was surpassed only by farmers' own quickening interest, with the result that the word *hybrid* itself became almost a synonym for high yield, rather than a specific reference to a genetic form. No notice was taken of the hundreds of individual lines, which differed as much from one another as hybrid differed from open-pollinated. This sense that hybrid corn was a singular type, that it was one of two corn alternatives, simplified a troublingly intricate situation, flattening out the very real complexity so farmers could consider their alternatives with abstract equanimity.[31]

[30]Jenkins, "Corn Improvement," p. 468.

[31]The farm press devoted much more attention to hybrids than the popular press; nevertheless, the range of popular interest can be seen in such articles as Andrew M. House, "Corn Belt Revolution," *Printer's Ink*, 15 December 1938, 11–14, 88–91; "Miracle Men of the Corn Belt," *Popular Mechanics*, August 1940, 226–229; "Hybrid Corn's Empire Grows," *Business Week*, January 1941, 28–30; and F. Thone, "Hybrid Corn's Conquests," *Science*, 26 April 1941, 271. Among farm publications, *Wallaces' Farmer* was unquestionably the primary forum for hybrid corn issues; see especially "The Story of Hybrid Corn," *Wallaces' Farmer*, 13 August 1938, 516–526.

On the other hand, actual hybrid corn choices were nothing short of alarming. By 1937 the University of Illinois College of Agriculture had developed one hundred hybrid combinations, and the Funk Brothers catalogue for 1938 listed thirty-six of the company's own hybrids plus twenty-nine station lines. Moreover, it was impossible for farmers to visually select the best hybrid because they were virtually identical in appearance. Farmers who had just the year before selected seed corn from their own fields, according to reliable university advice, who had tested it visually and on the germinator for quality, and who could hope to harvest a good crop that would again provide seed—such farmers must have felt small comfort reading Funk Brothers' 1935 catalogue: "You may not know which strain to order. Just order FUNK'S HYBRID CORN. We will supply the hybrid best adapted to your locality. . . . We want you to have the best." H. E. Klinefelter, writing for the *Missouri Farmer*, concurred: "The results we obtain from them [hybrids] depends upon the hybrids we plant, and this in turn depends upon the seed corn company where we buy our seed. We have to take their word for it—as to kind, quality, adaptability, etc. There are shysters in the hybrid seed corn business, just as there are in any other kind of business, and one has to be careful about where one buys seed. The thing to do in this respect, it would seem, is to buy seed from the big reliable companies whose investment in plants, advertising, and research work indicates that they expect to be permanent institutions."[32]

Although switching from open-pollinates to hybrids was a risky proposition for most farmers, properly adapted hybrids held an irresistible allure for many. Without question the most prominent feature of hybrid corn was its greater yield. Corn per-

[32]H. E. Klinefelter, "The Coming Revolution in Corn Production," *Missouri Farmer*, 15 October 1938, 5; University of Illinois Agricultural Experiment Station, "Illinois Cooperative Corn Improvement Program—1937 Policy for the Distribution of Available Supplies of Foundation Single Crosses and Inbred Lines of Corn" (hereafter "1937 Policy"), n.d., winter 1937, 8/6/2, box 5, UIA; FBSC, 1938 catalogue, 1935 catalogue, p. 1.

formance tests in 1934 and 1935 concluded that hybrids yielded on average over fifteen bushels more an acre than open-pollinated corn. Even though not all farmers could hope for such high yields because of adaptation problems, the dramatic contrasts were enough incentive for many to switch.[33]

Even in areas where the yield of hybrid corn was not much better than that of open-pollinated, hybrids often exhibited other features that appealed to farmers. For instance, most hybrids, because of their more extensive root systems and stiffer stalks, were not so easily flattened by high winds, and most farmers considered this "standability" nearly as important as yield. Corn that was knocked over by wind was more likely to rot on the ground before harvest, significantly reducing yield. Furthermore, "down" corn had to be harvested by hand under backbreaking conditions, unless it was simply turned over to livestock. Not coincidentally, farmers who had invested in the new corn-picking machinery could more easily justify the expense of hybrid corn, since the machinery was not able to pick corn off the ground. Some hybrids were more tolerant of cool weather than comparable open-pollinates, and some seemed more resistant to diseases and insects.[34]

But while hybrid corn enthusiasts, most notably industry spokespeople, have suggested that the characteristics of hybrids alone were responsible for the abandonment of open-pollinates, such a one-dimensional interpretation ignores two other important factors that were probably more influential in the short term.

[33]G. H. Dungan, J. R. Holbert, and A. L. Lang, "Progress in Hybrid Corn Production," (address delivered before the 41st Annual Meeting of the Illinois Farmers' Institute, February 1936) (Bloomington, Ill.: Pantagraph/Illinois Farmers' Institute, 1936), p. 33. Supporters of hybrids never tired of calculating the increased profit farmers could gain by growing hybrids; see, e.g., "Is Hybrid Seed Corn Profitable?" *Missouri Farmer*, 15 November 1938, 4; George Dungan et al., "Illinois Corn Performance Tests—Results for 1934," *Illinois AES Bulletin* 411, February 1935.

[34]H. W. Mumford, "Annual Report for 1936" (8/1/2, box 120, UIA), pp. 1, 7; Hackleman, "Annual Report for 1937," pp. 17, 25; Thomas Henley, farm adviser, report included in Hackleman, "Annual Report for 1939," p. 29.

The first was drought. In 1934 and again in 1936, Illinois farmers were beset by the deadly combination of too little rain and too much heat. Conditions in both summers were severe enough to reduce corn yield drastically, in turn reducing the amount of seed available for planting.[35] The scarcity of seed during these years generated increased interest in hybrids simply because any available seed—hybrid or open-pollinated—was better than nothing.[36]

The second major incentive for switching to hybrid corn was the acreage-reduction program of the Agricultural Adjustment Administration. As part of the corn-hog adjustment project, participating farmers reduced the amount of farmland planted to corn by 20 to 30 percent, using the land instead for pasture, for planting soil-improving crops, or for unharvested crops that would be turned over to livestock. In return, the government paid farmers thirty cents per bushel of corn not grown. Although another requirement of participation was that the farmer not increase total production of any marketable commodity, the AAA program was sufficiently complicated and difficult to comprehend that hybrid corn was apparently not considered a meaningful factor.[37]

[35]The census estimated that corn production dropped from 2,398 million bushels in 1933 to 1,449 million bushels in 1934. In 1935, which was a good year in terms of weather, the production figure rose to 2,299 million bushels, then fell again in 1936 to 1,506 million bushels. The figures for 1934 and 1936 were the lowest for the century. *Historical Statistics of the United States* (Washington, D.C: Bureau of the Census, 1976), p. 511. Although Lawrence H. Shaw and Donald D. Durost maintain that the weather had a negligible effect on overall yields from 1929 to 1962, their records indicate that in 1934 and 1936 weather distinctly reduced yields. See *The Effect of Weather and Technology on Corn Yields in the Corn Belt, 1929–1962* (1965).

[36]Hackleman, "Annual Report for 1935," p. 15. In his 1936 report, Hackleman noted that the scarcity of seed had created a "mild panic" for hybrids (pp. 49–50).

[37]See, for example, Theodore Saloutos, *The American Farmer and the New Deal* (1982), esp. pp. 72–75; Gilbert Fite, *American Farmers: The New Minority* (1981), pp. 71–72. Historians have not yet carefully examined the interaction between the AAA and hybrid corn, although it seems likely that hybrid corn was an important factor in the general failure of the AAA to reduce overall corn production significantly.

Because hybrid corn yielded more than open-pollinated, clever farmers found that they could be good citizens by taking land out of production without in fact reducing production. That is, farmers whose open-pollinated corn yielded sixty bushels an acre could switch to a hybrid that yielded eighty bushels an acre and reduce their acreage from ten acres to eight acres without sacrificing overall production. And these farmers, moreover, would receive payment from the government for cooperating in the reduction program. In this way many farmers circumvented the planned reduction by following the letter but not the spirit of the law. Beginning in 1932, national corn acreage did begin a gradual decline while the corresponding production figures either remained constant or increased slightly.[38]

Considering the variety of reasons for farmers to adopt hybrid corn, it is difficult to believe that any one was sufficiently compelling to provoke so widespread and abrupt a change. Rather, the rate at which farmers adopted hybrids with objective virtues seems to have been accelerated by economic considerations created by the seed shortage and AAA restrictions. That farmers who began using hybrids did not then switch back to open-pollinates when the economic crisis subsided in no way indicates that the field advantages of hybrids were solely responsible for the initial switch.

The Politics of Regulation

The seed corn shortage in 1936 prompted the college to start releasing its single-cross seed to farmers and growers who were otherwise unable to procure seed for planting. Both inexperienced and experienced growers were given seed, as well as

[38]In 1932 corn acreage harvested reached a peak of 110,577,000 acres; by 1940 it was down to 86,429,000 and by 1950 81,818,000. See also Hackleman, "Annual Report for 1940," p. 12; and Klinefelter, "The Coming Revolution," p. 5; both note the coincidence of AAA compliance and increased usage of hybrid corn.

elaborate instructions on how to tend these fields and detassel the corn. In cooperation with the Illinois Crop Improvement Association, the college repeated and in fact expanded the distribution in 1937, offering growers their choice of twenty-five different single-cross lines, twelve inbred lines developed in Illinois, and eleven inbreds developed at other stations. There were some restrictions: inexperienced growers, or "new producers," received only enough seed to plant one acre; experienced producers got enough to plant five acres; and all growers were required to have their fields inspected by the ICIA.[39]

The ICIA was a critical component in the college's strategy to control the production and distribution of hybrid seed in Illinois, which by 1936 had become exceedingly complicated. On one level, the college wanted to ensure that seed it had developed was kept free from contamination, and it used the ICIA as an enforcing agency that could threaten the small growers with noncertification of their corn. On this level the ICIA functioned as it always had for the college. On another level, the college sought to regulate the introduction and distribution of lines by the private breeders. Only the large companies were developing inbreds, and it seems clear that the college was trying to keep them in line.

Further, in 1936 the ICIA devised a new set of rules for certifying hybrid corn by which it attempted to extend the stringent growing and record-keeping practices of the college to private breeders. Under the new regulations, new hybrids could not be certified unless the inbred lines of which they were composed had also been certified. To be eligible for certification, hybrids had to "at least equal the average of the best five adapted open-pollinated varieties in lodging resistance and sound corn yield and exceed the weighted average performance rating of the best five adapted open-pollinated varieties by at least ten points." Producers were allowed to conduct the tests themselves under the supervision and inspection of the ICIA, but many chose to use

[39]Hackleman, "Annual Report for 1936," pp. 50–52; C. M. Woodworth to corn growers, 1 July 1936, included in ibid., pp. 76–77; "1937 Policy."

the university's corn performance tests both because the procedure was simpler and because the tests provided good free advertising. Another innovation was the separate designation of experimental and commercial hybrids. Experimental hybrids were new combinations that were not yet ready for commercial production, either because seed supplies were insufficient or because they had not yet demonstrated consistent superiority. In the first year of the three-year test, the hybrid could be entered as experimental and then switched to commercial. The commercial category was reserved for hybrids available on the market for sale.[40]

Certification of hybrids was a tricky business from the start and in many ways alienated the large producers. First, as an agent of the college, the ICIA was largely composed of small producers who depended upon university inbreds and single crosses to produce hybrids. When such producers received lines from the college, they had already been certified and the breeders lost no time in getting them into production. They needed only to maintain the lines satisfactorily and according to ICIA rules to obtain hybrid certification. In effect, such producers became agents of the college, producing and selling college or USDA hybrids from lines developed by them. Moreover, because they were dealing with public lines, that is, those developed by state or federal agencies, the small producers and the college used an open-pedigree system. Anyone who was interested could learn, for example, that Illinois 570, a station hybrid, was made up of USDA inbreds A, 90, 540, and Illinois experiment station inbred 540.[41]

Only one commercial producer—Lester Pfister—supported the open-pedigree system. According to Pfister, "if we have the right by law to know what goes into a 35-cent can of fly spray, we also have the right by common demand to know what inbreds and inbred crosses have gone into the production of the hybrid seed

[40]Hackleman, "Annual Report for 1935," pp. 35, 83–89; Dungan et al., "Progress," p. 11; Hackleman, "Annual Report for 1936," pp. 36–41.

[41]"1937 Policy."

corn for which we pay six to eight dollars a bushel—and on which we stake the greater part of our farm income." Pfister not only listed the pedigrees of his hybrids but also identified the source of the most widely used inbreds in commercial production generally as well as the six single-crosses that, he claimed, were in "90% of the highest yielding hybrid combinations."[42] Other large companies kept their pedigrees secret, even from their agents. One of Funk Brothers' associate producers complained, "I believe the research department has been too close with their information about the makeup of the actual hybrids we are now producing. It is not material to me, or to those in our company, what the actual inbreds are that go into the makeup of a given hybrid. I don't particularly care to know what [*sic*] R4 and Hy make up one of the single crosses in a particular number. But I do feel that we are entirely too ignorant as to the relation between certain hybrids; the reason why they may be drought resistant, grasshopper resistant, or chinch bug resistant. I feel that I should be able to talk to a customer and tell him that Funk's inbred 21, or single-cross 112, has certain characteristics, and that this single cross is contained in such and such hybrids."[43]

If the "middle managers" at Funk Brothers did not know the pedigrees of company lines, the ICIA was not likely to learn them either. While most large producers used one or two of their own inbred lines in their hybrids, they also used those developed by state and federal breeders and were not eager to divulge what they considered trade secrets. In order to receive certification, however, they were required to let ICIA inspectors check their fields for purity and seed identification, which amounted to reporting the pedigree of each combination. Their fear was that other producers could somehow obtain the one or two private lines, combine them with the widely available public lines, and then sell a hybrid identical to their own. This was a genuine possibility of

[42]*Pfister Hybrid Corn News*, Summer 1939, 2; Pfister, "Know Inbreds, Know Pedigrees," *Pfister Hybrid Corn News*, Fall 1938, 4.

[43]R. D. Herrington (J. C. Robinson Seed Company, Waterloo, Nebraska) to E. D. Funk, 13 December 1941, Funk Files.

course, but it would be naïve to suppose that the private breeders did indeed market distinct hybrid combinations. Because most inbreds developed in the early years in Illinois derived from the Krug strain of open-pollinated corn and because the private breeders traded inbred lines among themselves as well as used state and federal lines, it seems probable, if not certain, that some hybrids sold under different names were in fact identical. Hackleman noted in 1938 that US13 and Funk's G94 "seemed nearly identical," as did US5, G32, and DeKalb 817, and Illinois 960 and DeKalb 870. Nevertheless, because of these concerns, the large seed companies began refusing to certify their hybrid seed, or they certified only a fraction of it.[44]

If the nonparticipation of the large companies was distressing to the ICIA, it was sometimes devastating to small producers who operated as their agents and growers and who had not yet developed an independent reputation sufficient to sell hybrid seed without certification. For example, C. D. Ford, a grower who had obtained an inbred from Pfister, which he then used in a new hybrid, was denied certification because Pfister's inbred had not been certified. The rules could also be perplexing. After DeKalb bought some noncertified inbreds from Pfister, the company increased its supply, had them inspected by the ICIA, and made them available as certified inbreds.[45]

This almost farcical situation was exacerbated by the close resemblance of different hybrid strains. Open-pollinated corn was varied. In the early days of certification an experienced in-

[44]Hackleman, "Annual Report for 1938," p. 20. It was fairly easy to obtain inbred and single-cross lines simply by going into a competitor's field and taking a few ears. While the thief could probably not identify the strain, neither could anyone else. As to the similarity between hybrid lines with different names and numbers, one retired ICIA official whom I interviewed, when asked how many distinct inbred lines had been developed to date (1983), swept his arm to indicate the hundreds of files containing secret pedigrees and laughed, responding that these represented perhaps a few dozen lines.

[45]C. M. Woodworth to C. D. Ford, 22 April 1936, J. H. Roberts to Lloyd Pearson, 12 October 1937, both reproduced in Lester Pfister to H. P. Rusk, 30 April 1941, 8/1/2, Pfister Hybrid Corn Company, UIA.

spector could stand in a cornfield and tell at a glance if the corn was Krug or Plowman because different open-pollinated strains had different kernel shape, ear length, color, degree of luster, amount of foliage, and so on. Few, if any, inspectors could identify inbred lines on sight because they were so genetically similar that visual differences were nil. Thus the inspector relied heavily on the word of the producers; only they could identify a particular strain through their records.

By 1938, only two years after the new ICIA regulations, the college had all but lost its battle to control hybrid development in Illinois.[46] Meeting with the ICIA in early 1938, the Farm Crops Advisory Committee announced its conclusion that

> as the number of inbred lines increases, new lines are developed and old lines are back-crossed and changed and it is impossible for any inspector to go into single-crossing plots and be positive of his identification. Therefore, the larger companies who have their own plant breeders could be accepted as sources of foundation seed without actually inspecting the single crossing in the hands of these breeders. It was pointed out, however, that single-crossing plots in the hands of the individual farmers who obtained their seed either from the University or the Illini Corn Hybrids will be applying for inspection and this inspection should be provided.[47]

It was as if the two groups had agreed that hybrid inspection had become a ritualistic, rather than a regulatory device, useful only to the small producers who were still required to undergo it.

It is not difficult to understand why the college and the ICIA came to this decision, particularly in view of the membership of the two groups. The presence of Funk Brothers on the committees

[46]In 1937, the college and ICIA announced that they would no longer provide single-cross foundation seed to farmers; they helped establish Illini Corn Hybrids as a semiprivate organization that would henceforth manage the distribution of such seed.

[47]Meeting of Advisory Committee, University of Illinois Agronomy Department, 22 April 1938, 8/6/2, box 9, UIA.

was especially notable: Eugene Funk was a longtime member of the advisory board; O. J. Sommer, a Funk Brothers associate grower and first president of the ICIA, was also on the committee; and Earl Sieveking, who was a corn breeder at Funk Brothers, served in 1938 as vice-president of the ICIA. In some respects, the distinctions among the large companies, the college, and ICIA were superficial; in fact, by virtue of their participation in policy-oriented affairs, the large companies, and especially Funk Brothers, were apparently engaged in what amounted to self-regulation. This rather incestuous set of relationships was not lost on one group of farmers, who several months earlier had complained that "there seems to be a trend to undermine confidence in certification and a tendency for large commercial producers to get control of the pure lines to the detriment of all Illinois farmers."[48]

The large seed producers were disturbed by the distribution of inbred seed, which, of course, included lines they used in their hybrid combinations. At a meeting of the Farm Crops Advisory Committee and the agronomy Department staff in November 1938, the growing conflict between the college and the seed producers began to take shape. Breeders Eugene Funk and O. J. Sommer protested that the college was distributing inbreds to farmers who were incapable of keeping the lines pure. Funk asked "how many of those men have the natural ability and intuition to make good plant breeders," and Harold Lazier worried that the practice would "discredit the whole hybrid program." Sommer provided a chilling anecdote that, while probably exaggerated, expressed the worst fears of both public and private breeders: "Two years ago, a young man came up to me and said 'Do you know I had a letter from the University of Illinois, Crop Improvement Association. They asked me if I want to grow some pure lines and make some hybrid corn.' He then said to me, 'What is this hybrid corn? Do you think I could do the work—pulling off the tassels, growing it in rows, etc.?' I took the time and explained to him the best I

[48]Ford County Crop Improvement Association to ICIA, 19 January 1938, 8/6/2, box 9, UIA.

could. 'That sounds interesting,' he said, 'I think I will try it.' My reaction to that was that there was something wrong right then." Although everyone present knew that such an example was by no means typical, and probably responded to the story with nervous laughter, they realized as well that it was not an impossible scenario.[49]

Woodworth and Burlison became defensive, insisting in effect that their public service responsibility obliged them to accept the inherent riskiness of the project. They tried to make the committee understand the difficulty of their position:

> Woodworth: I don't believe you folks realize the pressure we were under in distributing this corn.
> Burlison: How do we know we are not going to pick a boy who will be a brilliant person in this work in a few years?
> Woodworth: We cannot arbitrarily choose the ones who should receive this seed for we cannot show partiality.
> Burlison: The USDA will send out seed to whoever asks for it.
> Woodworth: They send out 25 seeds and no questions asked. So we felt that a similar policy would be satisfactory for us to follow. In a public institution we cannot favor one and not the other.

As the college retreated from the regulation of hybrids, skirmishes with the companies continued, primarily over the corn performance tests. These tests were ultimately the only arena the two groups shared; the companies used the tests conducted by the college to advertise their new lines to farmers, and the college used the tests to show farmers the differences among the ever-increasing number of hybrids on the market. Not surprisingly, the tests provided yet another forum for dispute. Funk Brothers complained that the other big companies did not follow the test rules, charging DeKalb with changing the numbers on its lines and Pfister with putting identical lines in both the experimental and commercial divisions. Sieveking also complained about the final

[49]Advisory Committee and Staff, Agronomy Department Meeting, 1 November 1938, Alexander Collection, p. 15.

performance tabulations the college devised to summarize the tests, calling them "statistically incorrect and grossly misleading."[50]

But the college had complaints of its own. Hackleman reported that

> the performance tests do not contain hybrids that are most commonly sold in the area. The company or producer sometimes is more interested in getting a new hybrid into the test than in having an old one retested. The result is that the farmer may be getting the results of tests in which the hybrids which appear have no commercial import to him at the time. Still another objection is the fact that certain of the companies insert hybrids in the test which they are not growing in any quantity at the time and have no intention apparently of growing but which are high yielders and, therefore, serve a fine purpose as advertising matter.[51]

Hackleman was also annoyed by the closed pedigrees of the large breeders, which, he felt, put the college at a disadvantage. The college publicly identified the inbreds going into its own hybrids, and the farmer could identify by name or number those inbreds that contributed, for example, to a poor hybrid. Even though the large companies often used the identical inbred in their own lines, the farmer could not ascertain this information because company pedigrees were unknown. Furthermore, whereas the companies could point out particular inbreds in station lines to farmers, the station could not do the same with company lines. It became more and more apparent that the college was using the performance tests to evaluate new lines of unknown value, while the companies were using them to advertise lines that had already done well in company tests.[52]

[50]Earl Sieveking and J. R. Holbert to George Dungan, 4 February 1938, Sieveking to W. L. Burlison, 7 March 1940, J. C. Hackleman to Burlison, 30 April 1940, L. H. Smith to Burlison, 30 April 1940, and Burlison to Sieveking, 1 May 1940, Funk Files.

[51]Hackleman, "Annual Report, 1940," p. 20.

[52]Hackleman, "Annual Report for 1941," p. 7.

The increasingly haughty attitude of the large producers and their continual attempts to control hybrid policy in Illinois did not always work to their advantage. For example, Sieveking quarreled with college officials about the amount of hybrid corn the college required producers to enter in the commercial class of the performance tests. Sieveking felt that the five-bushel requirement was far too small and not in line with requirements of other states: Indiana required ten bushels and Iowa required fifty. Arguing that it was difficult to distinguish between the experimental and commercial classes with only five-bushel entries, he suggested that the college increase the requirement to one hundred bushels.[53]

Whatever his stated rationale, however, Sieveking's real object was fairly obvious. Since none but the largest producers had sufficient supplies of seed to enter such large quantities in the performance tests, the effect of the one-hundred-bushel requirement would be to eliminate the small producers from the tests and thus to deny them the exceptional advertising opportunities the tests provided. Further, since most small producers were testing station and federal lines, such a maneuver would practically eliminate the college as a seed producer in its own right. In the end Burlison dismissed Sieveking's suggestion for these very reasons.[54]

By 1942 the mechanics of distributing hybrid lines and negotiating policy had become distressing, tedious, and burdensome, not only for the University of Illinois but also for other midwestern land-grant institutions. Throughout the Midwest, university officials faced the same problems in deciding how best to distribute new lines to small and large producers and how to trade lines from one state to another. Interstate distribution was especially difficult because policies on certification, distribution to private breeders, and nomenclature differed substantially from one state to another, making it hard for stations to maintain control of an inbred line once it was released. At a 1937 meeting of

[53]Earl Sieveking to George Dungan, 8 March 1938, Funk Files.

[54]Burlison to Sieveking, 19 February 1940, Funk Files.

station and extension workers in corn-belt states the range of variability in hybrid corn policy became painfully evident. The Wisconsin station, for instance, would give new lines to commercial growers and stations in other states but not to Wisconsin growers. Although it is unclear why station officials pursued this policy, it is clear that Wisconsin growers could procure such lines from growers in other states. Wisconsin also wanted to change the hybrid name on lines released in other states; this strategy would not only create confusion over pedigrees but would challenge requirements that new lines be certified for three years even though the line, under its old name, had completed this trial. The certification process itself was also a problem, primarily because the cost varied widely from state to state. Thus commercial producers could choose where to grow seed based on a state's price policies. At least in theory, this price differential threatened hybrid swamping in some states and hybrid shortages in others.[55]

In an effort to systematize and standardize hybrid corn policy among these states, station directors convened again in 1941 to devise a distribution plan that would protect stations and their lines once seed was released to other states. Under the "delayed-release" policy, stations agreed to release inbred lines only after they had demonstrated definite value and combining ability and only after they had been increased sufficiently to meet demand. Each state then developed an organizational mechanism to handle distribution.[56]

The large producers were adamantly opposed to this arrangement, and called a meeting with members of the Agronomy Department to argue their case. They predicted that the program "would result in failure of cooperation between large producers and agricultural experiment stations," presumably because the regional delayed-release policy prevented the Illinois companies

[55]Minutes, Conference of Research and Extension Workers Interested in Corn Investigations, 12 March 1937, 8/1/2, box 32, file 32, UIA. Those in attendance represented Illinois, North Dakota, South Dakota, Michigan, Wisconsin, Minnesota, Kansas, Iowa, Nebraska, Indiana, and the Bureau of Plant Industry.

[56]Ibid. See also Hayes, *Professor's Story of Hybrid Corn*, pp. 188–201.

from obtaining new lines before small producers. In other words it seriously reduced their competitive advantage, and they suggested that the station should release lines as soon as such lines were shown to have value, not waiting until they had been increased in quantity and successfully combined with other lines. Ultimately this exception to delayed release—distributing lines that demonstrated superiority for a particular characteristic even though they had not yet been combined successfully—was allowed, probably because of effective lobbying by the large companies. In 1942 the large producers fought delayed release vigorously, and voted against the idea in a bloc. A representative of Pioneer Hi-Bred even went so far as to suggest that voting should be done according to acreage, with the votes of large producers counting for more than votes of small producers. This sort of bullying had become all too familiar to both college officials and small producers; the large companies were overruled, and college officials created the Illinois Seed Producers' Association. The ISPA, composed mostly of small producers, was designated "sole agent in the propagation of unreleased lines" developed by the Illinois station. The ISPA would buy the inbred seed from the college for a nominal fee, grow and multiply the seed, obtain certification, and sell the single-cross seed to commercial growers.[57]

In conjunction with this development, perhaps because they were emboldened by even such a modest coup, college officials decided to change certification requirements so that only open-pedigree hybrids would be eligible for the performance tests. In other words, if large producers insisted on keeping closed ped-

[57]Dean Rusk to hybrid corn producers (George Pfeifer, W. E. Riegel, C. B. Shuman, W. J. Mumm, J. R. Holbert, Roy Burrus, Porter Fox, L. L. Lowe, Merle Morgan, Hugh Morrison, Charles Nicholas), 6 March 1942. The arrangement was hammered out with producers between 14 March and 22 April. See "Progress Report on the Adoption of Program for Delayed Release of Inbred Lines of Corn in Illinois," prepared by C. M. Woodworth, 25 April 1942. Both the report and the Rusk letter can be found in Subject File, Independent Hybrid Producers, Inc., box 76, 1942, UIA.

igrees, then the ICIA would not accept their hybrids for testing and would restrict the tests to open-pedigree—that is, station and federal—hybrids. Holbert was not amused. Writing to George Dungan, he accused the college of trying to boost the sale and use of university lines beyond the merit of the seed. "Funk Brothers Seed Company," he sputtered, ". . . together with all the other major hybrid seed companies, cannot afford to have their reputation and the reputation of their products depend on a few tests conducted by farm advisers and busy farmers, few of whom are trained for this type of work and both are already overloaded with pressing duties and other obligations." Dungan replied that the object of the new rule was indeed to force the large producers either to cooperate with the college or to forfeit their advertising privilege. He declared that "from the standpoint of research," the closed pedigrees were of no use to anyone but the company. In the end an uneasy compromise was reached: the large companies certified a fraction of their hybrids under open pedigrees, just enough to keep their names in the farmer's eye.[58]

Experimentation to Commerce

One must beware of drawing too sharp a distinction between an experimental period and a commercial one. Nevertheless, by 1936 the most compelling and complex issues for those involved in hybrid corn centered on the rapid commercialization of hybrids. With the emergence of DeKalb and Pfister in 1935, both farmers and the college were confronted with questions for which there were no ready answers. Farmers were issued conflicting directives: seed companies urged them to buy hybrids, increase their incomes, and become scientific; the extension service urged them to act cautiously and continue using open-polli-

[58]J. R. Holbert to George Dungan, 6 April 1942, Dungan to Holbert, 9 April 1942, 8/6/2, subject file 99—Funk Brothers Seed Company, UIA.

nates; and the Agronomy Department seemed to encourage switching to hybrids by providing growers with seed for crossing and increase. Furthermore, farmers who did attempt to switch had to learn new field practices as well as the differences among hybrids, and they had to learn quickly or court economic disaster.

For the college the questions were both mechanical and ideological: how should inbred lines developed at state expense be distributed and to whom; what advice should be provided to farmers; how could farmers be protected from loss and educated in the new practices during the transition to hybrids; what sort of cooperation should obtain between private and public breeders; how could hybrid development be regulated so that the unbridled enthusiasm of the large companies could be tamed without infringing upon their rights and, indeed, their initiative; and most important by 1942, how could the college extricate itself from the hybrid fray without appearing irresponsible to farmers?

In creating the Illinois Seed Producers' Association, the college responded to the situation in characteristic fashion; distancing itself from the heat of the battle, it continued to determine overall policy and strategy. Throughout its history the college responded to crises by appointing semiautonomous associations that would follow college policy while representing a broader constituency; the Illinois Corn Growers' Association, the Farm Bureaus, the Illinois Agricultural Association, the Illinois Crop Improvement Association, and the Illinois Seed Producers' Association can all be viewed as mechanisms of college authority, which were created in response to, rather than in anticipation of, problems that grew out of the ambiguities in the relationship between the farmer and the college. In subsequent years the college did gradually leave hybrid corn to the major seed producers. George Sprague reportedly explained that the experiment stations "were not going to waste any time developing new inbreds in corn, that the hybrid seed corn research program was being so well handled by private companies [that] they were going to spend their time in other work. That is essentially true for all land-grant colleges."[59]

[59]George Sprague, quoted by J. R. Holbert in his testimony for J. A. Batson vs. J. C. Robinson Seed Company, 1956, Funk Files, pp. 80–81.

Whether Sprague's analysis is sufficient or whether the colleges simply tired of the struggle is perhaps immaterial. It is clear, however, that they negotiated themselves out of the picture by the mid-1940s, retreating from a contest they could neither win nor lose.

Conclusion:

Achieving Scientific Authority

In his address before the hybrid corn division at the annual American Seed Trade Association meeting in 1955, Henry A. Wallace reflected on the difference between the public and private development of hybrid corn. There had been, said Wallace, a natural division of labor between the experiment stations and USDA, on the one hand, and the private seed companies, on the other. The division was between pure and applied science. Whereas the public sector had been responsible for theoretical developments in corn breeding, the private sector had made advances in production-related areas. "In the minds of the majority of the Public," Wallace claimed, "including even better than average informed farmers, all corn breeding is being done by the private sector." Yet the hybrids that were now so widespread were, he continued, "one of the outstanding examples of close and practical cooperation between private individuals, the state experiment stations, and the USDA." But Wallace's emphasis on the merger of pure and applied science is neither very useful nor very apt in describing hybrid corn development. While there was a measure of cooperation between public and private institutions, clearly competition played an equal role. In this context competition refers not to commercial or even intellectual wrangling but to an intangible dispute over which kind of institution could claim scientific authority in the broader agricultural arena.[1]

[1]Wallace, "Public and Private Contributions."

Both the agricultural experiment station and agribusiness derived their strength from an ability to appear authoritative on agricultural matters. For the station, authority implied two things: first, the staff's expertise on a broad range of agricultural topics, from dairy herds to orchard pests to marketing, from abstract mathematical formulations to basic farm accounting methods; second, their impartiality in judging the assets and liabilities of agricultural innovations, their interpretation of rumors and advertisements. For agribusiness, authority implied integrity, an elusive concept usually involving a company's longevity, product reliability, and the perceived honesty of its owners. By the early twentieth century, authority also implied scientific methods for both the station and agribusiness. Scientific activity came to be seen as the basis of a station's expertise and impartiality, and for agribusiness it became a necessary ingredient of competition.

In the 1920s and 1930s three seed companies in Illinois linked their commercial authority and success with their scientific research programs in hybrid corn. DeKalb Agricultural Association, Pfister Hybrid Seed, and Funk Brothers all developed sophisticated programs similar to those undertaken by the experiment stations. They conducted hybrid research and studied corollary farm practices, invented mechanical farm equipment, compiled facts and figures attesting to their achievements, compared different strains, made recommendations, and in general bombarded farmers with scientific explanations for every puzzle.

Both the public and private groups were strategically involved in the early hybrid research and frequently cooperated on projects, shared experimental findings, drew on one another's expertise, and publicly commended one another's work. But they also competed in ways that reflected the different goals each played in Illinois agriculture. As impartial expert and judge, the college was an advocate for Illinois farmers, teaching them how to farm in a manner consistent with their needs and interests. In exchange for this attentiveness to their concerns, farmers deferred to the college, relying on its judgment on technical matters. The seed companies, however, had no such obligation to farmers.

Their relationship was more direct, relying on mutual satisfaction in a marketplace transaction.

This difference in roles became more pronounced when hybrid corn development shifted from the experimental to the commercial arena. While the college taught traditional methods of corn improvement, seed companies urged farmers to switch to hybrids. When the companies began conducting university-style research on hybrids, their role and authority in Illinois agriculture was enlarged. No longer just seed houses, they became dispensers of scientific knowledge. With their expanded scientific authority, the seed companies destabilized the college's own balance of authority, calling into question not only their relationship with each other but the proper role each should play.

While their research reflected differing roles, their behavior was not so distinctive. Indeed, taking Funk Brothers as the primary commercial actor in Illinois, the similarities between public and private researchers are striking. Both had been studying corn breeding and nomenclature since at least the 1890s. At the Illinois station chemist Cyril G. Hopkins invented the ear-to-row method of selection, which allowed for easy identification of parents, and demonstrated that the chemical composition of corn could be altered by selecting for particular characteristics, such as high or low oil or protein content. These studies were of little direct benefit to farmers but helped identify the commercial limits of breeding. Perry Holden and A. D. Shamel began inbreeding studies in 1895 in an effort to understand inheritance patterns, and in the first decade of the twentieth century L. H. Smith studied hybrid vigor and the transmission of unit characters in corn, in a fairly typical scientific response to the rediscovery of Mendel's work in 1900. And in the 1920s and 1930s, George Dungan and C. M. Woodworth examined inheritance of characteristics, particularly chemical composition of hybrids.[2] Funk, the commercial breeder, was also active. In 1892 he developed a

[2]D. E. Alexander, "Early Work in Corn Experimentation"; University of Illinois, Department of Agronomy, "Historical Data for President Kinley," February 1941, 8/6/2, UIA.

specialized varietal cross called Funk's 90-Day designed for late planting or short growing seasons. In that year he also began using Hopkins's ear-to-row method of corn improvement by which he developed smooth, Utility-type corn, creating a minor revolution that ultimately redefined standards for open-pollinated corn.[3]

Both groups were involved in such Illinois farmers' organizations as the Illinois Corn Breeders' Association, which Dean Eugene Davenport created in 1901 and in which Eugene Funk was a prominent member and officeholder. Funk also sat on the Soils and Crops Advisory Committee at the college. Finally, while one expects college agronomists to be well acquainted with contemporary scientific trends, it is notable that Eugene Funk was also a participant in the scientific network, corresponding with early geneticists such as Donald Jones and George Shull and hosting Hugo De Vries on his tour of American agricultural centers.

But the differences between Funk Brothers and the college more readily account for their divergent concerns in corn research and development. First, despite the college's impressive array of scientists, technicians, laboratories, and equipment, its commitment to hybrid corn was mediated by the institution's multidisciplinary nature. Corn farmers and producers were not the only clientele clamoring for attention, and the college and station had other priorities. Soil surveys occupied most of the research time at the station, followed closely by cultural studies of new crops such as alfalfa, clover, and soybeans. Nor did the college possess the experimental fields necessary for full-scale corn research. In the 1910s its available farmland was both sparse and of marginal quality.

Funk Brothers, on the other hand, owned over twenty-two thousand acres of prime Illinois farmland, constituting an enormous field laboratory. And while the company continued to produce and sell open-pollinated corn as well as other crop seed, its interest in hybrids was complete and uncompromising. Funk

[3] J. D. Funk, "Agronomist's Report," 2 May 1903, Funk Files; Holbert, "Funk's 176-A Story."

Brothers' research projects on corn extended beyond the standard inbreeding studies and gave the company the appearance of a hybrid corn experiment station unfettered by the usual demands of farmers. Its catalogues from the late 1920s on were testimonials to scientific experimentation and research, describing in detail the procedures used in hybridizing. The company also sponsored its own yield contests, modeled on those of the experiment station, for farmers growing Funk Brothers' strains.

The most fundamental difference between the two groups, however, was reflected in their very different organizational structures, which provided distinct opportunities as well as constraints on the kind of science each could develop. At Funk Brothers, the commercially oriented research program was given a substantial boost by long-term support from the USDA. During the twenty years the station was in operation, it accelerated the pace of hybrid production more than any other single farm or station, and by Funk Brothers' own admission, distinguishing the contributions of the company from those of the USDA was virtually impossible. While the USDA did not aid the company much in terms of capital and equipment, it did provide it with unusual advantages. Most important, the USDA supported Holbert, who by the 1930s was easily one of the most sophisticated and experienced corn breeders in the country. The federal link also allowed access to the research of other breeders, both public and private, the results of which were not really available to small-scale breeders. And of course the prestige Funk Brothers gained from the USDA connection undoubtedly enhanced its business and commercial integrity and was useful in attracting academic scientists such as Earl Sieveking, who in 1928 left the University of Illinois for a research position at Funk Brothers. Not surprisingly, college scientists and administrators were less than thrilled with federal stations like the one at Funk Brothers; in 1931 an Alabama experiment station employee called them "one of the most embarrassing features of federal-state relations."[4]

[4]Leon Steele, interview with author, 1984; M. J. Funchess et al., "Report of Committee on Federal-State Relations in Agricultural Research," 17 November 1931, 8/1/2, box 83, UIA.

The college also had a formal relationship with the USDA through federal appropriations, but its corn work commanded no special attention at the federal level, and in fact relations between Davenport and the federal government were strained at best. While Funk was negotiating with federal officials to establish the field station, Dean Davenport at the college was canceling a cooperative agreement with the USDA, creating a tension that undoubtedly contributed to the decision to exclude the college from connection with the field station in Illinois.[5] The research and extension efforts of the college were most clearly shaped by its commitment to the needs and interests of Illinois farmers and the college's own political interests. While college scientists conducted a significant amount of theoretical research and published in professional journals, the experiment station publications tended to concentrate on routine farm issues. In corn research especially, practical work was dominant. Between 1903 and 1928 the station undertook a dozen projects aimed at systematizing farm practices, compared with four concerned with hybrid research, two of which were cooperative projects instigated by Funk Brothers.

The practical corn studies were of two types: those aimed at systematizing traditional farming techniques and those focused on systematizing traditional methods of corn improvement, that is, identifying and selecting good seed from the field for planting. When the Division of Crops Extension was established in 1922, its director, J. C. Hackleman, concentrated on teaching farmers how to visually distinguish between good and bad seed corn, how to increase their yields through such selection, and how to build and use seed germinators. He also encouraged them to enter the annual statewide ten-acre corn yield contest sponsored by the college. Hackleman emphasized the extension ideal of self-help, teaching techniques that would enable farmers to manage their farms more productively and scientifically. In addition, he cre-

[5]See Davenport, "Recommendation for the Abrogation of Cooperative Relations with the USDA," 2 November 1916, Subject File, box 83, UIA. This action led to the cancellation of the memo of 1914, which established federal-state relations in Illinois.

ated the Illinois Crop Improvement Association, whose goals of identifying seed types and increasing supplies of pure seed were achieved through an elaborate system of inspection and certification. For a time, the ICIA enjoyed a certain popularity among pure-seed enthusiasts and grew in membership and number of acres certified, but its excessively complicated requirements for certification alienated the largest seed companies. Although company executives sat on the ICIA board of directors, the large seed houses stopped certifying their own seed. The college ultimately waived its requirement that entrants in the yield contest certify their entries, providing they had a corn breeder on staff, thereby reluctantly acknowledging that the college could not control large companies' research or success.

By the mid-1930s, when hybrid corn became commercially available, the college and Funk Brothers had built up momentum in quite different directions, which were reflected in their approach to extending results to farmers. Funk Brothers, eager to get its hybrids on the market before the competitors, claimed, more enthusiastically than accurately, that it had produced the first commercial hybrid as early as 1916. Actually, the company introduced its first true double-cross hybrids in the 1928 catalogue, prefacing the particular descriptions with a lengthy discussion of scientific corn breeding. While Funk Brothers did not recommend that farmers plant only the new hybrids and, in fact, charged double the price of premium open-pollinates, its enthusiasm for hybrids was entirely uncritical. The company broke down resistance among farmers with such incentives as a seed package containing both hybrid and open-pollinated corn, free hybrid seed for buyers of open-pollinates, and even free hybrid seed to boys entering the yield contest in exchange for names and addresses of farmers. Although Funk Brothers continued to sell open-pollinates until 1940, such corn received less and less attention in the catalogue.

These persuasive techniques would be of little use, however, if Funk Brothers could not surmount the true obstacle to acceptance: regional adaptation. In 1929 the company shipped nearly a thousand bushels of hybrid seed to growers in thirty-five states, Canada, and Cuba. Much to its chagrin, all of it failed except the corn grown in Bloomington. Moreover, even though most hybrid corn yielded at a higher rate than open-pollinates in a good year, a year of drought, early frosts, late rains, chinch bugs, rootworms, or any other natural disaster would claim the specialized hybrids and leave the more "experienced" open-pollinates untouched. Research and development could not end with the first hybrid. By 1939 Funk Brothers offered over thirty of its own specialized hybrids plus nearly thirty hybrids developed by agricultural colleges.

The college, faced with commercial hybrid corn, reacted with conservatism, caution, and frustration. First, though college research contributed substantially to hybrid development, extension efforts had successfully promoted a system of farm practice fundamentally incongruent with hybrids. For example, although farmers could modestly improve their open-pollinates by selecting good seed from the field as Hackleman taught them, such a practice had disastrous consequences with hybrid corn. Since second-generation hybrids were substantially less productive than first-generation ones, farmers who used the hybrids two years in a row suffered a much-decreased yield. Farmers repeatedly tried selecting seed for planting from their hybrid fields, either unable or unwilling to believe that they needed to buy new seed every year. In addition, in the early years when available hybrids were not adapted to specific conditions, a farmers' local open-pollinate would often perform better than hybrids. This problem in adaptation led to many exaggerated claims both for and against hybrids, leaving experts at both Funk Brothers and the college at a loss to predict the performance of any single strain.

Second, college officials were uncertain how to proceed with their own inbreds and single-cross seed. In 1931 it seemed plausible for farmers to do their own hybridizing with college inbreds; by 1940 the idea was all but dead. Once the college began disseminating inbreds for crossing, it became difficult to ensure that participating farmers bred them properly and responsibly. Some farmers received inbreds and single-cross seed and sold them as true hybrids, confusing everyone and reflecting badly on the college. By 1937 the college had developed a program in conjunction with the ICIA for distributing inbreds and single-cross seed which catered to established seed producers. The large producers obtained station inbreds for crossing with their own lines, and the program also made a serious attempt to supply the smaller seed houses that did not have breeding programs and were therefore entirely dependent on the station.

By the late 1930s the hybrid corn juggernaut had begun to roll. Whereas in 1933 hybrids accounted for only 0.4 percent of corn acreage in the United States, by 1945 they made up 90 percent. Certainly, the virtues of hybrids—higher yield and resistance to environmental problems—offered persuasive reasons for switching from open-pollinates. Nevertheless, there were two more compelling incentives. First, the adoption rate was hastened by several bad crop years, especially the drought in 1936, when seed of all kinds became scarce. Not only were farmers eager to obtain high-yielding seed corn during a period of high prices, but they were also receptive to advertisements describing hybrid corn bred especially for drought resistance. In addition, the Agricultural Adjustment Administration launched its acreage-reduction program in 1934, whereby farmers were encouraged to take land out of production. For many farmers, the prospect of reduced acreage, steady output, and government payments was the ultimate incentive to try hybrids, whose higher yield per acre allowed farmers to produce as much as ever, if not more.

For the college, the success of hybrid corn signaled the end of an era. Private agricultural research centers like that at Funk Brothers were claiming a share of the scientific authority formerly monopolized by land-grant institutions and the USDA. Moreover, the college's role as farmers' advocate became tangled in conflicts between tradition and innovation, public and private, caution and enthusiasm. As advocate, the college attempted to "put a brake on" early adoption of hybrids when it became clear that adaptation was incomplete, and through the ICIA, it tried to regulate the distribution of station inbreds in a way that gave small producers a chance to compete. The college also continued sponsoring and publishing the results of state corn yield contests, which attempted to evaluate fairly the assets and liabilities of different hybrids throughout the region.

Despite these efforts, however, the concentration of authority, if not expertise, shifted unmistakably to commercial firms. Farmers could no longer select their own seed; they were obliged to follow the seed company's advice on appropriate seed. When in doubt, Funk reassured farmers, let us choose the corn that is best for you. And indeed, all hybrids looked virtually identical, making visual selection impossible. Only those who understood the mysterious pedigree system could make such important decisions. By 1938 the college was advising farmers to rely on the integrity of big seed companies. The college was no longer the sole proprietor of agricultural expertise, and in fact private centers could often claim broader familiarity with both practical farming and research trends. But while some argued that the college should concentrate on theoretical science and leave practical matters to private researchers, the college's public service function demanded a more activist role. It should come as no surprise, then, that in the wake of hybrid corn, the college shifted its attention from hybrid corn production to corn marketing and by-product conversion, issues the private firms were content to ignore.

What does this story tell us about the differences between public and private research in agriculture and about the relationship between pure and applied science? If the behavior of farmers and scientists in one state can be considered indicative of broader national patterns, and I think it can, then several claims are warranted. First, this particular half century was characterized by land-grant agriculturalists' uncertainty about their proper role in the states and their proper relationship with farmers. Farmers' acceptance hinged in large part on the degree to which the land-grant university could respond to farmers' needs and interests, which were ultimately economic in character. Farmers everywhere wanted to reduce farming costs, increase prices for crops and livestock, make fieldwork less arduous, and so on. They were particularly attentive to ideas and techniques that could facilitate these goals, and college scientists, though by no means abandoning the more abstract issues of agricultural science, did attempt to solve farm problems and explicitly linked their work with the greater economic health of the farm population.

Hybrid corn symbolized a conjunction of diverse interests: scientists' interest in testing Mendelian theory, farmers' interest in increased yield and profit and better field characteristics, and seed producers' interest in predictable and profitable corn. I suspect that such conjunctions are not at all uncommon in agricultural development. Looking at the history of other accepted innovations, such as agricultural implements or chemical agents, I would be surprised to find a different scenario. In many ways it would be more instructive to locate innovations that failed to win the acceptance of the farming, commercial, or land-grant sector, for then we could see more clearly how the assumptions one group held about another proved false and on what issues the ensuing dispute turned.

Second, the story of hybrid corn broadens our understanding of how the land-grant university became involved with agribusiness. Some recent analysts, such as Jim Hightower, have suggested that the land-grant institutions are little more than handmaidens of agribusiness, providing companies with expertise and research that is neither scientifically nor socially justifiable.[6] It is important to consider how this relationship developed, what alternatives existed for the university at each juncture, and why it chose the role it did. While I am in no way an apologist for land-grant behavior, I think that further examination will demonstrate the complexities and compromises that such institutions contended with in carving out a role in American science and American agriculture.

This case of hybrid development raises thorny questions about the role of the land-grant system in commodity production and its responsibility to those engaged in agriculture as farmers and

[6]Hightower, *Hard Tomatoes, Hard Times.*

seed producers. Clearly, hybrid corn was not a uniform benefit to all. Large seed companies were enormously strengthened, while smaller seed producers were often absorbed by large producers or squeezed out of the market entirely. The benefit to farmers, moreover, has been mixed. Hybrids bred to withstand specific adverse field conditions have made corn growing a more stable and predictable venture, but the social and economic costs have been considerable. Not only must farmers buy new seed each year, but hybrid corn introduced an array of corollary farm products such as fertilizer, insecticides, herbicides, and other pesticides; the equipment used to apply these chemicals; and the enormous (and enormously expensive) machinery used to plant and harvest corn. In the past thirty years such additional "inputs" have attained the status of farming necessities for all but a tiny minority of agriculturalists. Further, the higher yielding capacity of hybrid corn, which initially increased farmers' production and income, has had the overall effect of sustaining chronic overproduction and declining farm prices.

For the land-grant university, this history of hybrid corn provides a useful lesson in the politics of agricultural research. At issue is how the university identifies its primary responsibilities and clientele. As we have seen, the Illinois agricultural population represented a broad spectrum of needs and interests and benefited unevenly from hybrid development. As Jim Hightower has pointed out, the increasing interdependence of agribusiness and the land-grant university often has a negative impact on farmers and consumers and undermines the "good faith" understanding between the farmer and the university by which the university is perceived as a public servant acting in the best interests of the farmer. Yet the relationship among farmers, commercial seed producers, and the university has always been ambiguous, and the research-and-development choices made by the university have rarely been guided by long-term policy commitments. Whether this ambiguity has worked to the advantage of American agriculture is unclear. In the case of hybrid corn, it seems that only the seed companies emerged from the turmoil unbruised.

Bibliography

Notes on Primary Sources

Materials relating to the University of Illinois College of Agriculture were located in the University Archives. Two exceptions were the reports of the Crops Extension Specialist from 1922 to 1942, which were in the Office of the Crops Extension Specialist, and a small collection of correspondence and reports in the possession of D. E. Alexander, Department of Maize Genetics.

Funk Brothers has preserved a sizable collection of historical material, consisting mainly of correspondence, clippings, copies of addresses delivered by the staff, and a few reports from the agronomist. Two documents were especially useful: the Application for Relief under Section 722 of the Internal Revenue Code for the Fiscal Years Ended June 30, 1941, 1942, and 1943 contained not only a history of the company but tables and graphs charting incremental stages of commercial growth; and J. R. Holbert's testimony at the trial of *J. A. Batson vs. J. C. Robinson Seed Company* provided insights into the nature of agricultural research and development. The annual spring catalogues put out by the company were quite helpful as well. Correspondence between J. R. Holbert and USDA officials pertaining to the federal field station at Funk Farms is located in the National Archives in Washington, D.C.

Pfister Hybrids had destroyed pertinent records only months

before I requested them, but did provide a nearly complete run of their two company publications: *Pfister Hybrid Corn News*, the annual catalogue, and *Pfister Flashes*, which was designed for salesmen.

In addition to a small collection of correspondence and the transcription of a lengthy interview with Charles Gunn, DeKalb Ag Research records included an inventory of its two publications, the *Winged Ear* for farmers and *Acres of Gold* for salespeople. In addition their sales manual, like Funk's application for relief, provided an unexpected but quite helpful view into the scientific and commercial components of research and development at DeKalb.

Finally, the records of the Bureau of Plant Industry and the Office of Cereal Investigations, located in the National Archives, were extremely useful for understanding the bureau's shift in corn research and recommendations in the 1920s, as well as the influence of Wallace and Jones in bureau affairs.

Sources

Aldrich, Darragh. *The Story of John Deere: A Saga of American Industry.* Minneapolis: McGill, 1942.

Alexander, D. E. "Early Work in Corn Experimentation at Illinois." Manuscript, 1959, Alexander Collection, Department of Maize Genetics, University of Illinois.

——. "Illinois and the Beginnings of Hybrid Corn." Manuscript, 1963, Alexander Collection, Department of Maize Genetics, University of Illinois.

Allen, E. W. "Position and Outlook of the Experiment Stations." *Proceedings, Association of American Agricultural Colleges and Experiment Stations*, 1919, 128–136.

Allen, Garland. "Hugo De Vries and the Reception of the Mutation Theory." *Journal of the History of Biology*, 1969, 2:55–87.

——. *Thomas Hunt Morgan.* Princeton: Princeton University Press, 1978.

Bibliography

Babcock, Ernest B., and R. E. Clausen. *Genetics in Relation to Agriculture*. New York: McGraw-Hill, 1918.

Babcock, Jarvis J. "Adoption of Hybrid Corn: A Comment." *Rural Sociology*, 1962, 27:332–338.

Bailey, Joseph C. *Seaman A. Knapp: Schoolmaster of American Agriculture*. New York: Columbia University Press, 1945.

Baker, Gladys. *The County Agent*. Chicago: University of Chicago Press, 1939.

Beal, William J. "Reports." Michigan State Board of Agriculture, 1876, 1877, 1881, 1882.

Becker, Stanley L. "Donald F. Jones and Hybrid Corn." *Bulletin of the Connecticut Agricultural Experiment Station*, 1976, 763:1–9.

Berlan, Jean-Pierre. "Heterosis and Hybrid Corn: Much Ado about Nothing?" Manuscript, in possession of the author, 1986.

Berlan, Jean-Pierre, and Richard Lewontin. "The Political Economy of Hybrid Corn." *Monthly Review*, 1986, 38:35–47.

Block, William J. *The Separation of the Farm Bureau from the Extension Service*. Urbana: University of Illinois Press, 1960.

Bogue, Allan G. *From Prairie to Corn Belt: Farming on the Illinois and Iowa Prairies in the Nineteenth Century*. Chicago: University of Chicago Press, 1963.

Broehl, Wayne G., Jr. *John Deere's Company: A History of Deere and Company and Its Times*. Garden City, N.Y.: Doubleday, 1984.

Brown, William J. "H. A. Wallace and the Development of Hybrid Corn." *Annals of Iowa*, 1983, 47:167–179.

Bruce, A. B. "The Mendelian Theory of Heredity and the Augmentation of Vigor." *Science*, 1910, 32:627–628.

Bryant, T. R., et al. "Report of the Special Committee to Study Types of Extension Organization and Policy in the Land-Grant Colleges." *Assoc. Am. Agri. Coll. Exp. Sta.*, 1914, 260–262.

Burritt, M. C. *The County Agent and the Farm Bureau*. New York: Harcourt, Brace, 1922.

Busch, Lawrence, and William B. Lacy. *Science, Agriculture, and the Politics of Research*. Boulder, Colo.: Westview Press, 1983.

Campbell, Christiana McFadden. *The Farm Bureau and the New Deal*. Urbana: University of Illinois Press, 1962.

Cannon, Grant, ed. *Farm Quarterly Great Men of Modern Agriculture*. New York: Macmillan, 1963.

Carnegie Foundation for the Advancement of Teaching. *Fourth Annual Report*. Boston: Merrymount, 1909.

Carriel, Mary Turner. *The Life of Jonathan Baldwin Turner*. Urbana: University of Illinois Press, 1961.

Carrier, Lyman. "A Reason for the Contradictory Results in Corn Experiments." *Journal of the American Society of Agronomy*, 1919, 2:106–113.

Carstensen, Vernon. "The Genesis of an Agricultural Experiment Station." *Agricultural History*, 1960, 34:13–20.

Castle, W. E. "The Beginnings of Mendelism in America." In L. C. Dunn, ed., *Genetics in the Twentieth Century*. New York: Macmillan, 1951, pp. 59–76.

——. "Recent Discoveries in Heredity and Their Bearing on Animal Breeding." *Proceedings of the American Breeders' Association*, 1905, 2:120–126.

Cavenagh, Helen M. *Funk of Funk's Grove: Farmer, Legislator, and Cattle King of the Old Northwest, 1797–1865*. Bloomington, Ill.: Pantagraph, 1952.

——. *Seed, Soil, and Science: The Story of Eugene D. Funk*. Chicago: Lakeside Press, 1959.

Churchill, Frederick B. "Wilhelm Johannsen and the Genotype Concept." *Journal of the History of Biology*, 1974, 1:5–30.

Collins, G. N. "Dominance and Vigor of First Generation Hybrids." *American Naturalist*, 1921, 55:116–133.

——. "Notes on the Agricultural History of Maize." *Agricultural History*, 1923, 2:411–429.

——. "The Value of First-Generation Hybrids in Corn." *United States Bureau of Plant Industry Bulletin* 191, 1910, 7–40.

Conover, Milton. *The Office of Experiment Stations: Its History, Activities, and Organization*. Baltimore: Johns Hopkins University Press, 1924.

Crabb, A. Richard. *The Hybrid Corn Makers: Prophets of Plenty*. New Brunswick, N.J.: Rutgers University Press, 1948.

Daniel, Pete. *Breaking the Land: The Transformation of Cotton, Tobacco, and Rice Cultures since 1880*. Urbana: University of Illinois Press, 1985.

Darwin, Charles. *The Effects of Cross and Self Fertilization in the Vegetable Kingdom*. London: John Murray, 1876.

Davenport, Eugene. "The American Agricultural College." *Assoc. Am. Agri. Coll. Exp. Sta.*, 1912, 159.

——. "Rejuvenation of the College of Agriculture of the University of Illinois, 1895–1922." Manuscript, November 1933, University of Illinois Archives.

——. "Report of the Committee on Experiment Station Organization and Policy." *Assoc. Am. Agri. Coll. Exp. Sta.*, 1915, 123–125.

——. "Shall We Ask for Further Legislation in the Interests of Agriculture? If So, What?" *Assoc. Am. Agri. Coll. Exp. Sta.*, 1910, 79.

De Kruif, Paul. *The Hunger Fighters*. New York: Blue Ribbon Books, 1928.

De Vries, Hugo. *Intracellular Pangenesis*, trans. C. S. Jaeger. Chicago: Open Court, 1910.

——. *The Mutation Theory*, trans. J. C. Farmer and A. D. Darbishire. Chicago: Open Court, 1910.

——. *Plant Breeding*. Chicago: Open Court; London: Kegan Paul, 1907.

Dickson, J. G., and J. R. Holbert. "Relation of Temperature to the Development of Disease in Plants." *American Naturalist*, 1928, 62:311–333.

"Don't Guess about Seed Corn." *Prairie Farmer*, 3 February 1930, 16–17.

Douglass, Benjamin Wallace. "A Revolution in Corn Growing." *Country Gentleman*, 1920, 5:47.

Dungan, G. H., J. R. Holbert, and A. L. Lang. "Progress in Hybrid Corn Production." Bloomington, Ill.: Pantagraph/Illinois Farmers' Institute, 1936.

Dungan, G. H., et al. "Illinois Corn Performance Tests—Results for 1934." *Illinois AES Bulletin* 411, February 1935.

East, Edward Murray. "The Distinction between Development and Heredity in Breeding." *American Naturalist*, 1909, 43:173–181.

——. "Genotype Hypothesis in Maize." *American Naturalist*, 1911, 45: 160–174.

——. "Heterosis." *Genetics*, 1936, 21:375–397.

——. "The Improvement of Corn in Connecticut." *Conn. AES Bulletin* 152, 1906.

——. "The Relation of Certain Biological Principles to Plant Breeding." *Conn. AES Bulletin* 158, 1907.

——. "Report of the Agronomist: I. The Prospects for Better Seed Corn in Connecticut." *Conn. AES Report*, 1907–1908, 397–405.

——. "Report of the Agronomist: II. Practical Use of Mendelism in Corn Breeding." *Conn. AES Report*, 1907–1908, 406–418.

——. "Report of the Agronomist: III. Inbreeding in Corn." *Conn. AES Report*, 1907–1908, 419–428.

——. "Role of Hybridization in Plant Breeding." *Popular Science Monthly*, 1910, 77:342–355.

East, Edward Murray, and H. K. Hayes. "Heterozygosis in Evolution and Plant Breeding." *USDA Bureau Plant Industry Bulletin* 243, 1912, 8–48.

——. "Inheritance in Maize." *Conn. AES Bulletin* 167, 1911.

East, Edward Murray, and Donald F. Jones. *Inbreeding and Outbreeding*. Philadelphia: Lippincott, 1919.

Emerson, R. A. "Bean Breeding." *Proc. American Breeders' Association*, 1903, 1:50–55.

——. "Inheritance of a Recurring Somatic Variation in Variegated Ears of Maize." *Nebraska AES Research Bulletin* 4, 1913, 5–35.

Ettling, John. *The Germ of Laziness*. Cambridge: Harvard University Press, 1981.

"Farmers Learn Seed Testing." *Prairie Farmer*, 12 February 1931, 18.

Fite, Gilbert. *American Farmers: The New Minority*. Bloomington: Indiana University Press, 1981.

Fitzharris, Joseph C. "Science for the Farmers: The Development of the Minnesota Agricultural Experiment Station, 1868–1910." *Agricultural History*, 1974, 48:202–214.

Fosdick, Raymond B. *Adventure in Giving: The Story of the General Education Board*. New York: Harper and Row, 1962.

Fraser, Colin. *Tractor Pioneer: The Life of Harry Ferguson*. Athens: Ohio University Press, 1973.

Froggart, P., and N. C. Nevin. "The 'Law of Ancestral Heredity' and the Mendelian-Ancestrian Controversy in England, 1889–1906." *Journal of Medical Genetics*, 1971, 8:1–36.

Fuller, F. E. "Science Hits the Corn Belt." *Country Gentleman*, September 1939, 9.

Funk, E. D. "Commercial Corn Breeding." Address presented to the Congress of Experiment Stations and Colleges of Agriculture, Louisiana Purchase Exhibition, 8 October 1904, Funk Files.

——. "The Search for Better Corn." Manuscript for an address in St. Charles, Illinois, 28 October 1938, Funk Files, (never delivered).

Funk, J. Dwight. "Commercial Corn Breeding." *Proc. American Breeders' Association*, 1903, 1:29–33.

——. "Practical Corn Breeding on a Large Scale." *Proc. American Breeders' Association*, 1905, 3:89–93.

Galloway, Beverly T. "Farm Demonstration and Farm Management Work of the Bureau of Plant Industry of the USDA." *Assoc. Am. Agri. Coll. Exp. Sta.*, 1911, 97–98.

——. "Organization of Cooperative Extension Work: Machinery and Method (in the State)," *Assoc. Am. Agri. Coll. Exp. Sta.*, 1915, 220–231.

Garrison, H. S., and F. D. Richey. "Effects of Continuous Selection for Ear Type in Corn." *USDA Bulletin* 1341, 1925.

General Education Board. *The General Education Board: An Account of Its Activities.* New York: General Education Board, 1915.

Glover, Wilbur H. *Farm and College: The College of Agriculture of the University of Wisconsin: A History.* Madison: University of Wisconsin Press, 1952.

Gowan, J. W., ed. *Heterosis.* Ames: Iowa State University Press, 1952.

Griffee, Fred. "First Generation Corn Varietal Crosses." *Journal of the American Society of Agronomy*, 1922, 14:18–27.

Griliches, Zvi. "Hybrid Corn and the Economics of Innovation." *Science*, 1960, 132:275–280.

——. "Hybrid Corn: An Explanation of the Economics of Technological Change." *Econometrics*, 1957, 25:501–522.

——. "Profitability versus Interaction: Another False Dichotomy." *Rural Sociology*, 1962, 27:327–330.

——. "Research Costs and Social Returns: Hybrid Corn and Related Innovations." *Journal of Political Economy*, 1958, 66:419–431.

Gross, Neal C. "The Diffusion of a Culture Trait in Two Iowa Townships." Master's thesis, Iowa State University, 1942.

Gustavson, Reuben. *History Is Prologue: Land-Grant Centennial Program.* Urbana: University of Illinois Press, 1964.

Hackleman, J. C., compiler. *History [of the] International Crop Improvement Association, 1919–1961.* Clemson, S.C.: Chambers Printing, 1961.

Handschin, W. F. "The County Agent and the Farm Bureau." *Assoc. Am. Agri. Coll. Exp. Sta.*, 1919, 287–290.

Harding, T. Swann. *Two Blades of Grass: A History of Scientific Development in the USDA.* Norman: University of Oklahoma Press, 1947.

Harris, J. A. "Biometric Proof of the Pure-Line Theory." *American Naturalist*, 1911, 45:346–363.

Hartley, C. P. "Corn Breeding Work in the USDA." *Proc. American Breeders' Association*, 1903, 1:33–37.

——. "Plant Breeding Principles Applied to Corn Improvement." *Proc. American Breeders' Association*, 1905, 2:108–112.

——. "Progress in Methods of Producing Higher-yielding Strains of Corn." *Yearbook of Agriculture*, 1909, 309.

Hartley, C. P., E. B. Brown, G. H. Kyle, and L. L. Zook. "Crossbreeding Corn." *USDA Bureau Plant Industry Bulletin* 218, 1912.

Havens, Eugene, and Everett M. Rogers. "Adoption of Hybrid Corn: Profitability and the Interaction Effect." *Rural Sociology*, 1961, 25:409–414.

——. "Another False Dichotomy." *Rural Sociology*, 1962, 27:330–332.

Hayes, Herbert Kendall. "Present-Day Problems of Corn Breeding." *Journal of the American Society of Agronomy*, 1926, 18:344–363.

——. *Professor's Story of Hybrid Corn*. Minneapolis: Burgess, 1963.

Hayes, Herbert Kendall, and R. J. Garber. *Breeding Crop Plants*. New York: McGraw-Hill, 1927.

——. "Synthetic Production of High-Protein Corn in Relation to Breeding." *Journal of the American Society of Agronomy*, 1919, 11:308–318.

Hayes, Herbert Kendall, and F. R. Immer. *Methods of Plant Breeding*. New York: McGraw-Hill, 1942.

Hegleson, Robert C. "Hybrid Corn Finds Market in Seed Trade." *National Seedsman*, October 1938, 8–9.

Heitmann, John L. *The Modernization of the Louisiana Sugar Industry, 1830–1910*. Baton Rouge: Louisiana State University Press, 1987.

Hightower, Jim. *Hard Tomatoes, Hard Times—The Original Hightower Report, Unexpurgated, of the Agribusiness Accountability Project on the Failure of America's Land Grant College Complex*. Cambridge, Mass.: Schenkman, 1972.

Historical Statistics of the United States. Washington, D.C.: Bureau of the Census, 1976.

Hoffer, G. N., and J. R. Holbert. "Results of Corn Disease Investigations." *Science*, 1918, 47:246–247.

——. "Selection of Disease-Free Seed Corn." *Indiana AES Bulletin* 244, 1918.

Holbert, James R. "Control of Corn Rots by Seed Selection." *Ill. AES Circular* 243, 1920.

——. "Funk's 176-A Story." Manuscript, 1945, Funk Files.

——. "Germination Story." Manuscript, 1950, Funk Files.

——. "A Great Industry of One Decade." *Finance*, 25 September 1944, 34–36.

——. "Selection of Disease-Free Seed Corn." *Indiana AES Bulletin* 244, 1918.

——. "Thirty Years Breeding Corn." *Seed World*, 5 June 1942, 116–120.

Holbert, James R., W. L. Burlison, B. Koehler, C. M. Woodworth, and G. Dungan. "Corn Root, Stalk, and Ear Rot Diseases and Their Control through Selection and Breeding." *Ill. AES Bulletin* 255, 1924.

Holbert, James R., and G. N. Hoffer. "Control of the Root, Stalk, and Ear Rot Diseases of Corn." *USDA Farmers' Bulletin* 1176, 1920, 1–24.

——. "Corn Root and Stalk Rots." *Phytopathology*, 1920, 10:55.

Holbert, James R., P. E. Hoppe, and A. L. Smith. "Some Factors Affecting Infection with and Spread of *Diplodia* in the Host Tissue." *Phytopathology*, 1935, 25:1113–14.

Hopkins, Cyril G. "The Chemistry of the Corn Kernel." *Ill. AES Bulletin* 53, 1898.

——. "Experiments in Corn Breeding." *Proc. American Breeders' Association*, 1903, 1:65–68.

——. "Inbreeding of Corn and Methods of Prevention." *Proc. American Breeders' Association*, 1905, 2:147–150.

——. "The Soil as a Bank." Address to the Fourth Annual Convention of the Illinois Bankers' Association, 15 June 1910, University of Illinois Archives.

——. "The Story of a Corn King and Queen." *Popular Science Monthly*, March 1911, 251–257.

House, Andrew M. "Corn Belt Revolution." *Printer's Ink*, 15 December 1938, 11–14, 88–91.

"How Much Corn Can You Grow?" *Prairie Farmer*, 26 April 1930, 5.

"Hybrid Corn's Empire Grows." *Business Week*, January 1941, 28–30.

"Hybrid Corn Stands Up." *Prairie Farmer*, 12 April 1930, 30.

"Hybrid Corn Wins Boys' Corn Contest." *Wallaces' Farmer*, 21 February 1931, 249.

"Hybrid Corn Yields Well." *Prairie Farmer*, 8 March 1930, 14.

Iftner, George H. "Memorial to James L. Reid: Pioneer Corn Breeder." *Illinois State Historical Society Journal*, 1955, 48:427–430.

"Is Hybrid Seed Corn Profitable?" *Missouri Farmer*, 15 November 1938, 4.

James, Edmund J. *Sixteen Years at the University of Illinois, 1904–1920*. Urbana: University of Illinois Press, 1920.

Jenkins, Merle T. "Corn Improvement." *Yearbook of Agriculture*, 1936, 455–495.

——. "A New Method of Self-pollinating Corn." *Journal of Heredity*, 1923, 14:41–44.

Johannsen, W. "The Genotype Concept of Heredity." *American Naturalist*, 1911, 45:129–159.

Jones, Donald F. "Biographical Memoir of Edward Murray East." *National Academy of Sciences Biographical Memoirs*, 1944, 23:217–242.

——. "Dominance of Linked Factors as a Means of Accounting for Heterosis." *Genetics*, 1917, 2:466–479.

——. "The Effects of Inbreeding and Cross Breeding upon Development." *Conn. AES Bulletin* 207, 1918, 5–100.

——. "Inbreeding in Corn Improvement." *Breeders' Gazette*, 1919, 75: 1111–1112.

——. "Selection in Self-fertilized Lines as a Basis for Corn Improvement." *Journal of the American Society of Agronomy*, 1920, 12:77–100.

Jones, Donald F., and Paul C. Manglesdorf. "Improvement of Naturally Cross-pollinated Plants by Selection in Self-fertilized Lines." *Conn. AES Bulletin* 266, 1925.

Jones, Donald F., and W. R. Singleton. "The Improvement of Naturally Cross-pollinated Plants by Selection in Self-fertilized Lines: II. The Testing and Utilization of Inbred Strains of Corn." *Conn. AES Bulletin* 375, 1935.

Jorgensen, Louis, and H. E. Brewbaker. "A Comparison of Selfed Lines of Corn and First Generation Crosses between Them." *Journal of the American Society of Agronomy*, 1927, 19:819–830.

Jugenheimer, Robert W. *Hybrid Maize Breeding and Seed Production*. Rome: FAO, 1958.

Kahn, E. J., Jr. *The Staffs of Life*. Boston: Little, Brown, 1985.

Keeble, F., and C. Pellow. "The Mode of Inheritance of Stature and the Time of Flowering in Peas." *Journal of Heredity*, 1910, 1:47–56.

Keller, Evelyn Fox. *A Feeling for the Organism: The Life and Work of Barbara McClintock*. San Francisco: Freeman, 1983.

Kent, George. "Bagging a Million." *Country Home Magazine*, August 1938, 7, 21.

Kerr, W. J. "Some Land-Grant College Problems." *Assoc. Am. Agri. Coll. Exp. Sta.*, 1911, 37–51.

Kevles, Daniel J. "Genetics in the United States and Great Britain, 1890–1930: A Review with Speculations." *Isis*, 1980, 71:441–455.

Kiesselbach, T. A. "A Half-Century of Corn Research, 1900–1950." *American Scientist*, 1951, 39:629–655.

Kimmelman, Barbara A. "The American Breeders' Association: Genetics and Eugenics in an Agricultural Context." *Social Studies of Science*, 1983, 13:163–204.

——. "Hugo De Vries and the Experimental Method." Manuscript, 1978, in author's possession.

——. "A Progressive Era Discipline: Genetics at American Agricultural Colleges and Experiment Stations, 1890–1920." Ph.D. diss., University of Pennsylvania, 1987.

Klinefelter, H. E. "The Coming Revolution in Corn Production." *Missouri Farmer*, 15 October 1938, 5.

Kloppenburg, Jack Ralph, Jr. *First the Seed: The Political Economy of Plant Biotechnology, 1492–2000*. Cambridge: Cambridge University Press, 1988.

Knapp, Bradford. "Relationship of Agricultural Extension Work to Farmers' Cooperative Buying and Selling Organizations." *Assoc. Am. Agri. Coll. Exp. Sta.*, 1917, 305–310.

Knoblauch, H. C., E. M. Law, and W. P. Myer. *State Agricultural Experiment Stations: A History of Research Policy and Procedures*. USDA Misc. Publ. 904, Washington, D.C., 1962.

Kommelade, W. G. "Sassafras and Persimmons: History of the Dixon Springs Agricultural Center, Illinois." *University of Illinois College of Agriculture, Cooperative Extension Service, Special Publication 40*, 1976, 38–139.

Kuhn, Thomas. *The Structure of Scientific Revolutions*. Chicago: University of Chicago Press, 1962.

Lacey, John J. *Farm Bureau in Illinois*. Bloomington, Ill.: Illinois Agricultural Association, 1965.

Lang, Alvin C. *Fifty Years of Service: A History of Seed Certification in Illinois, 1922–1972*. Urbana: Illinois Crop Improvement Association, 1973.

McCleur, George. "Corn Crossing." *Ill. AES Bulletin 21*, 1892, 82–101.

McConnell, Grant. *The Decline of Agrarian Democracy*. Berkeley: University of California Press, 1959.

Manglesdorf, Paul C. *Corn: Its Origin, Evolution, and Improvement*. Cambridge: Harvard University Press, 1974.

——. "Hybrid Corn." *Scientific American*, 1951, 185:39–47.

Marcus, Alan. *Agricultural Science and the Quest for Legitimacy*. Ames: Iowa State University Press, 1985.

——. "Comment" (on W. L. Brown, "H. A. Wallace and the Development of Hybrid Corn." *Annals of Iowa*, 1983, 47:167–169). Ibid., 180–189.

"Miracle Men of the Corn Belt." *Popular Mechanics*, August 1940, 226–229.

Moores, Richard Gordon. *Fields of Rich Toil: History of the University of Illinois College of Agriculture*. Urbana: University of Illinois Press, 1970.

Morrow, George, and Frank Gardner. "Field Experiments with Corn." *Ill. AES Bulletin* 24, 1893, 173–203.

Morrow, George, and Thomas Hunt. "Field Experiments with Corn." *Ill. AES Bulletin* 4, 1889, and *Bulletin* 13, 1891.

Mosher, Martin. *Early Iowa Corn Yield Tests and Related Later Programs*. Ames: Iowa State University Press, 1962.

Mumford, F. B. "Cooperation in Extension Work between the USDA and the Colleges of Agriculture." *Assoc. Am. Agri. Coll. Exp. Sta.*, 1912, 135–140.

——. *The Land-Grant College Movement*. Columbia: University of Missouri Press, 1940.

Mumford, H. W. "The College of Agriculture and the Farm Bureau." Manuscript, University of Illinois Archives, 1924.

——. "The Influence to Date of Smith-Lever Extension Work on Rural Life in the United States." *Assoc. Am. Agri. Coll. Exp. Sta.*, 1929, 256–263.

Mumm, W. J., and C. M. Woodworth. "Heritable Characters of Maize: 36-A Factor for Soft Starch in Dent Corn." *Journal of Heredity*, 1930, 21:503–505.

Neufeld, E. P. *A Global Corporation: A History of the International Development of Massey Ferguson, Limited*. Toronto: University of Toronto Press, 1969.

Nevins, Allan. *The Origins of Land-Grant Colleges and State Universities*. Washington, D.C.: GPO, 1962.

Nilsson-Leissner, Gunnar. "Relation of Selfed Strains of Corn to F_1

Crosses between Them." *Journal of the American Society of Agronomy*, 1927, 19:440–454.

Norman, Thomas. *Minneapolis-Moline: A History of Its Formation and Operations*. New York: Arno, 1976.

Nourse, Edwin G., Joseph S. Davies, and John D. Black. *Three Years of the Agricultural Adjustment Administration*. Washington, D.C.: Brookings Institution, 1937.

Oliver, Francis. "Success Saga of the Corn Belt." *St. Louis Post-Dispatch*, 22 September 1938, 1.

Payne, Walter A., ed. *Benjamin Holt: The Story of Caterpillar Tractor*. Holt and Atherton Pacific Center for Western Studies, Monograph Series 2:1. Stockton, Calif.: University of the Pacific, 1982.

Powell, Fred W. *The Bureau of Plant Industry: Its History, Activities, and Organization*. Baltimore: Johns Hopkins University Press, 1927.

"Produce Certified Seed." *Prairie Farmer*, 7 February 1931, 16.

Pugsley, C. W., et al. "Report of the Committee on Extension Organization and Policy." *Assoc. Am. Agri. Coll. Exp. Sta.*, 1916, 134–139.

Rankin, Fred H. "Work of the College of Agriculture." *Illinois Argricultural Progress*. Urbana: University of Illinois Press, 1922.

Rasmussen, Wayne D., ed. *Growth through Agricultural Progress, USDA Centennial Year*. Washington, D.C.: USDA, 1961.

Rasmussen, Wayne D., and G. L. Baker. *The Department of Agriculture*. New York: Praeger, 1972.

Rexroat, P. W. "Before the Beginning: History of the Dixon Springs Agricultural Center in Illinois." *University of Illinois College of Agriculture, Cooperative Extension Service Special Publication* 40, 1976, 1–35.

Rexroat, P. W., and L. S. Mayer. "The Productiveness of Successive Generations of Self-fertilized Lines of Corn and of Crosses between Them." *USDA Bulletin* 1354, 1925.

Rexroat, P. W., G. H. Stringfield, and G. F. Sprague. "The Loss of Yield That May Be Expected from Planting Second Generation Double-Crossed Seed Corn." *Journal of the American Society of Agronomy*, 1934, 26:196–199.

Richey, F. D. "Convergent Improvement of Selfed Lines of Corn." *American Naturalist*, 1927, 61:430–449.

——. "Effects of Selection on the Yield of a Cross between Varieties of Corn." *USDA Bulletin* 1209, 1924.

——. "The Experimental Basis of the Present Status of Corn Breeding." *Journal of the American Society of Agronomy*, 1922, 14:1–17.

Robbins, E. T. "The Farm Bureau." *Illinois Agricultural Policy*. Urbana: University of Illinois Press, 1922, pp. 71–78.

Roberts, H. F. *Plant Hybridization before Mendel*. Princeton: Princeton University Press, 1929.

Roepke, Howard G. "Changes in Corn Production on the Northern Margin of the Corn Belt." *Agricultural History*, 1959, 33:126–132.

Rosenberg, Charles E. "The Adams Act: Politics and the Cause of Scientific Research." *Agricultural History*, 1964, 38:3–12.

——. *No Other Gods: On Science and American Social Thought*. Baltimore: Johns Hopkins University Press, 1978.

——. "Rationalization and Reality in the Shaping of American Agricultural Research, 1875–1914." *Social Studies of Science*, 1977, 7:401–422.

——. "Science and Social Values in Nineteenth-Century America: A Case Study in the Growth of Scientific Institutions." In Arnold Thackray and Everett Mendelsohn, eds., *Science and Values: Patterns of Tradition and Change*. New York: Humanities Press, 1974, pp. 21–42.

——. "Science in American Society: A Generation of Historical Debate." *Isis*, 1983, 74:356–367.

——. "Science Pure and Science Applied: Two Studies in the Social Origin of Scientific Research." In Rosenberg, *No Other Gods: On Science and American Social Thought*. Baltimore: Johns Hopkins University Press, 1978, pp. 185–195.

——. "Science, Technology, and Economic Growth: The Case of the Agricultural Experiment Station Scientist, 1875–1914." *Agricultural History*, 1971, 45:1–20.

Ross, Earle D. *Democracy's College: The Land-Grant Movement in the Formative Stage*. Ames: Iowa State University Press, 1942.

Rossiter, Margaret. *The Emergence of Agricultural Science: Justus Liebig and the Americans, 1840–1880*. New Haven: Yale University Press, 1975.

——. "The Organization of Agricultural Improvement in the United States, 1785–1965." In Alexandra Oleson and S. C. Brown, eds., *The Pursuit of Knowledge in the Early American Republic: American Learned Societies from Colonial Times to the Civil War*. Baltimore: Johns Hopkins University Press, 1976, pp. 279–298.

——. "The Organization of the Agricultural Sciences." In John Voss and Alexandra Oleson, eds., *The Organization of Knowledge in America, 1860–1920*. Baltimore: Johns Hopkins University Press, 1979, pp. 211–248.

Ryan, Bryce. "A Study in Technological Diffusion." *Rural Sociology*, 1948, 13:273–285.

Ryan, Bryce, and Neal Gross. "The Diffusion of Hybrid Seed Corn in Two Iowa Communities." *Rural Sociology*, 1943, 8:15–24.

Saloutos, Theodore. *The American Farmer and the New Deal*. Ames: Iowa State University Press, 1982.

Sandster, Emil P. "Heredity in the Light of Recent Investigations." *Proc. American Breeders' Association*, 1905, 2:183–185.

Schapsmeier, Edward L., and Frederick H. Schapsmeier. *H. A. Wallace of Iowa: The Agrarian Years, 1910–1940*. Ames: Iowa State University Press, 1968.

Schlebecker, John T. *Whereby We Thrive: A History of American Farming, 1607–1972*. Ames: Iowa State University Press, 1974.

Scott, Roy V. *The Agrarian Movement in Illinois, 1880–1896*. Illinois Studies in the Social Sciences, volume 52. Urbana: University of Illinois Press, 1962.

——. *The Reluctant Farmer: The Rise of Agricultural Extension to 1914*. Urbana: University of Illinois Press, 1970.

Searle, A. M. "Administration of the Smith-Lever Act." *Assoc. Am. Agri. Coll. Exp. Sta.*, 1914, 119–129.

Shaw, Lawrence H., and Donald D. Durost. *The Effect of Weather and Technology on Corn Yields in the Corn Belt, 1929–1962*. USDA, Economic Research Service, 1965.

Shull, George. "The Composition of a Field of Maize." *Proc. American Breeders' Association*, 1908, 4:296–301.

——. "Experiments with Maize." *Botanical Gazette*, 1911, 52:480–485.

——. "The Genotypes of Maize." *American Naturalist*, 1911, 45:234–252.

——. "A Pure-Line Method of Corn Breeding." *Proc. American Breeders' Association*, 1909, 5:51–59.

Simmonds, Norman. *Principles of Crop Improvement*. New York: Longman, 1979.

Singleton, W. Ralph. "Early Researches in Maize Genetics." *Journal of Heredity*, 1935, 26:49–59, 121–126.

Smith, A. L., and J. R. Holbert. "Corn Stalk Rot and Ear Rot." *Phytopathology*, 1931, 21:129.

Smith, L. H. "The Effect of Selection on Certain Physical Characters in the Corn Plant." *Ill. AES Bulletin* 132, 1909.

Smith, L. H., and A. M. Brunson. "Experiments in Crossing Varieties as a Means of Improving Productiveness in Corn." *Ill. AES Bulletin* 306, 1928.

Sprague, George. "The Changing Role of the Private and Public Sectors in Corn Breeding." *35th Annual Corn-Sorghum Research Conference*. Washington, D.C.: American Seed Trade Association, 1980, 1–9.

——. *Corn and Corn Improvement*. New York: Academic Press, 1954.

——. "Hybrid Corn." *Yearbook of Agriculture*, 1962, 106–107.

Steece, Henry M. "Breeding Work with Field Crops at the Experiment Stations." *Office of Experiment Stations Annual Report*, 1924, 43–59.

Steele, Leon. "The Hybrid Corn Industry in the United States." In David B. Walden, ed., *Maize Breeding and Genetics*. New York: John Wiley and Sons, 1978, pp. 29–40.

Stevenson, John A. "Plants, Problems, and Personalities: The Genesis of the Bureau of Plant Industry." *Agricultural History*, 1954, 28:155–162.

"The Story of Hybrid Corn." *Wallaces' Farmer*, 13 August 1938, 516–526.

[Student], "The Probable Error of a Mean." *Biometrika*, 1908, 6:1–25.

Sturtevant, A. H. "The Early Mendelians." *Proceedings of the American Philosophical Society*, 1965, 109:199–204.

Taylor, W. A. "Relation of the Work of the Bureau of Plant Industry to Agricultural Extension." *Assoc. Am. Agri. Coll. Exp. Sta.*, 1912, 140–149.

Thompson, D. O. "The Illinois Agricultural Association." *Illinois Agricultural Policy*. Urbana: University of Illinois Press, 1922, pp. 79–81.

Thompson, W. O., et al. "Report of the Executive Committee." *Assoc. Am. Agri. Coll. Exp. Sta.*, 1915, 20–21.

Thone, F. "Hybrid Corn's Conquests." *Science*, 26 April 1941, 271.

True, A. C. "Administration of the Smith-Lever Exension Act." *Assoc. Am. Agri. Coll. Exp. Sta.*, 1914, 113–118.

——. *A History of Agricultural Experimentation and Research in the United States, 1607–1925*. USDA Misc. Publ. 251, 1937.

——. *A History of Agricultural Extension Work in the United States, 1785–1923*. USDA Misc. Publ. 15, 1929.

——. "Report of the Bibliographer: A History of the Hatch Experiment Station Act of 1887." *Assoc. Am. Agri. Coll. Exp. Sta.*, 1926, 95–107.

True, A. C., and V. A. Clark. *Agricultural Experiment Stations in the United States*. USDA, Office of Experiment Stations, Bulletin 80, 1900.

University of Illinois, Department of Agronomy. "Historical Data for President Kinley, February 1941." 8/6/2, University of Illinois Archives.

Wallace, Henry A. "Public and Private Contributions to Hybrid Corn, Past and Future." *Proceedings of the 10th Annual Hybrid Corn Industry Research Conference*. Washington, D.C.: American Seed Trade Association, 1955.

——. "The Revolution in Corn Breeding." *Prairie Farmer*, 21 March 1925, 1, 5.

Wallace, Henry A., and William L. Brown. *Corn and Its Early Fathers*. East Lansing: Michigan State University Press, 1956.

Watts, R. L. "Additional Federal Support for Experiment Station Work." *Assoc. Am. Agri. Coll. Exp. Sta.*, 1919, 253–255.

Webber, H. J. "Correlation of Characters in Plant Breeding." *Proc. American Breeders' Association*, 1906, 3:73–83.

——. "Explanation of Mendel's Law of Hybrids." *Proc. American Breeders' Association*, 1905, 2: 138–143.

Webber, H. J., and Ernst A. Bessey. "Progress of Plant Breeding in the United States." *Yearbook of Agriculture*, 1899, 465–490.

White, H. C. "The Experiment Stations." *Assoc. Am. Agri. Coll. Exp. Sta.*, 1912, 80–86.

"Who's Who in the Seed Trade—Eugene D. Funk." *Seed World*, 19 December 1924, 31.

Wiest, Edward A. M. *Agricultural Organization in the United States*. Lexington: University of Kentucky Press, 1923.

Wik, Reynold M. *Benjamin Holt and Caterpillar Tractors and Combines*. St. Joseph, Mich.: American Society of Agricultural Engineers, 1984.

Williams, C. G. "Corn Breeding and Registration." *Proc. American Breeders' Association*, 1907, 3:110–122.

Williams, Robert C. *Fordson, Farmall, and Poppin' Johnny: A History of the Farm Tractor and Its Impact on America*. Urbana: University of Illinois Press, 1987.

Winter, F. L. "The Effectiveness of Seed Corn Selection Based on Ear Characters." *Journal of the American Society of Agronomy*, 1925, 17:113–118.

Woodworth, C. M. "Heritable Characters of Maize: 28-Barren Sterile." *Journal of Heredity*, 1926, 17:405–411.

Zirkle, Conway. *Beginnings of Plant Hybridization*. Philadelphia: University of Pennsylvania Press; London: Oxford University Press, 1935.

Index

Index

Library of Congress Cataloging-in-Publication Data

Fitzgerald, Deborah Kay.
The business of breeding : hybrid corn in Illinois, 1890–1940 / Deborah Fitzgerald.
p. cm.
Includes bibliographical references.
ISBN 0-8014-2233-7 (alk. paper)
1. Corn—Illinois—Breeding. 2. Corn—Breeding. I. Title.
SB191.M2F645 1990
633.1'523'09773—dc20 89-46179